MEANS WITHOUT END

A Critical Survey of the Ideological Genealogy of
Technology without Limits, from Apollonian *Techne*
to Postmodern Technoculture

Gregory H. Davis

University Press of America,® Inc.
Lanham · Boulder · New York · Toronto · Oxford

Copyright © 2006 by
University Press of America,® Inc.
4501 Forbes Boulevard
Suite 200
Lanham, Maryland 20706
UPA Acquisitions Department (301) 459-3366

PO Box 317
Oxford
OX2 9RU, UK

Library of Congress Control Number: 2006924913
ISBN-13: 978-0-7618-3485-4 (paperback : alk. paper)
ISBN-10: 0-7618-3485-0 (paperback : alk. paper)

™
⊖ The paper used in this publication meets the minimum
requirements of American National Standard for Information
Sciences—Permanence of Paper for Printed Library Materials,
ANSI Z39.48—1984

Promethée, seul, est devenu dieu et regne sur la solitude des hommes.

Albert Camus,
L' Homme Revolté

Contents

Preface

When I published *Technology: Humanism or Nihilism* in 1981, the technological society of which Jacques Ellul had written was already a reality, although the arrival of *technoculture* was yet to come. What was clear, in any case, was that the phenomenon of technology without limits was an integral part of contemporary existence. My book, *Technology: Humanism or Nihilism*, had dealt with the impact of modern technology on many aspects of contemporary existence; but I was already beginning to search for an understanding, from a historical and cultural perspective, of how the phenomenon of technology without limits had emerged in the first place. The following year, as a Research Associate at the Office of the History of Science and Technology of the University of California, Berkeley, I decided to begin the project that eventually resulted in the completion of this book, *Means Without End*, after a number of delays and interruptions. In the process of pursuing this goal, I discovered that what I was investigating were not only the ideological origins of technology without limits, but the genesis of the technoculture it has infused and come to define.

My interest in modern technology and its contradictions was stimulated by a number of factors. On the level of personal experience, these included study at Stanford University, a particularly favorable venue for engineering and the idea of technical methods and solutions for essentially nontechnical matters; over thirty years of residence in Northern California's Silicon Valley, where technological innovation and promotion without end are part of the very air one breathes; a generation of teaching at the college level the course Technology, Contemporary Society, and Human Values; and two years' work as a political scientist at Stanford Research Institute dealing with the political and strategic aspects of nuclear weapons, a technology whose dangerous implications clearly placed the question of limits in the context of self-preservation.

On the level of books and ideas, I was particularly influenced by Ellul's *Technological Society*, which demonstrated that modern technology had become autonomous in practice; Hannah Arendt's description in her *Origins of Totalitarianism* of the inherent nihilism of an ideology whose ultimate goals are process and movement, regardless of the consequences; Albert Camus' *The Rebel*, one of the great books of the second half of the twentieth century, in which he lamented the fact that modern Western culture had nihilistically rejected the Greek idea of limits; Neil Postman's *Technopoly*, in which he maintained that information glut was subverting authentic knowledge, culture and values in the Information Age; and Paul Virilio's *Information Bomb*, where he argued that postmodern technology, of incredible power and sophistication, creates a dangerous false consciousness and transcends all ethical and practical limits, posing a grave threat to human well-being and existence. I should also mention that my studies of literature in France in the early 1960s contributed to the humanistic perspective that I have always applied to my analysis of technology and its influence on contemporary humans and their culture.

To write critically about technology is a solitary undertaking, not only because very few people do it, but also because everything functions in our postmodern technoculture, on both an ideological and a practical level, to promote the belief that technology and technological development without limits are unmitigated blessings. One should keep in mind, however, that my intention is not to deny technology its obvious importance and value in human existence, but rather to focus on the ideological transformations leading to the loss of philosophical and practical limits for it and the emergence of our present technoculture.

Since this work is essentially an examination of ideas about technology in the broader historical and cultural context of Western civilization in the last two and a half millennia, the greater part of my effort has been concentrated on finding linkages for a broader synthesis rather than intensive scholarship in one or several narrowly defined and highly specialized areas. I believe that in the age of the postmodern "little narrative," with its lack of any broader philosophical or historical perspective, this work may serve a useful purpose for students, technologists, academicians, and others concerned with technology and cultural issues. In pursuing my endeavor, I benefited greatly from discussion and interaction with a number of individuals. Some of these dialogues enriched my understanding of various issues; others opened doors to new and fruitful paths of investigation. Among these individuals were Ronald G. Smith, Dr. Ronald Puppo, Dan Kaplan, Michael Chriss, Waldo Esteva, Irving and Hela Norman, Dr. David Noble, Dr. Clement Kleinstreuer, Dr. Albert Acena, Dr. David Danielson, Isago Tanaka, Alonso Smith, Karen Dorst, Michael Makovsky,

Michael Kimball, Donald Porter, Chip Rouzie, Murray Tobak, and Jack Stauffacher.

I want to acknowledge a special debt for encouragement and advice from Dr. Martin Jay and Dr. Roger Hahn, both at the University of California, Berkeley; and in particular I would like to thank the latter for invaluable support during my fellowship there. I am likewise grateful to John Dooley, Barlow Weaver, Gladys Chaw, and Colette Norman, librarians who were of invaluable help on numerous occasions. Special thanks are in order for Cesar de Araujo, a software engineer of many talents, for practical assistance, critique, and encouragement at a crucial time for the success of this project. In providing moral support, Denise Flowerday, Clara Israel, and Sharon Brown all played important roles, not to mention Leonard Davis, whose zest for living was infectious and always a source of encouragement.

Gregory H. Davis
Berkeley, California
November, 2005

Introduction

The Phenomenon of Technology without Limits: A Defining Characteristic of Contemporary Culture

In the contemporary world, the traditional relation between technology and culture has been inverted. Instead of being a part of culture, subject to philosophical and practical limits, technology has essentially absorbed all aspects of life and culture into its own methods and purposes. The result is that ever-increasing numbers of human beings live a hypertechnologized existence in what can only be accurately termed a *technoculture*, a civilization of means dominated by the phenomenon of technology without limits.

In this technoculture, technology has taken on an exaggerated importance in human existence. Humans define their existence in terms of it and look to it as the means to achieve all their dreams and aspirations. They see technology as the embodiment of the superiority and greatness of their species, and consider likewise that the greatness of their nation and culture is undisputable if it includes a high level of technological development. Similarly, they expect technology to provide a solution to all their problems, although many of our most serious problems are now caused by technology itself. And in the face of mounting evidence that ever-expanding technology poses a threat to the existence of humans and other forms of life on the planet, humans strive nevertheless to enhance even more its power and sophistication.

In both theory and practice, modern technology has essentially become autonomous, independent of any valid semblance of ethical control. Modern science, once concerned only with understanding the phenomenal world, has become technoscience, expected always to yield practical applications. And despite the fact that technology is a human activity, in many respects humans are becoming its slaves rather than its masters. Modern technology has greatly magnified the power of humans vis-à-vis other humans and nature while authentic culture, necessary to guide technology and moderate its risks and dangers, has been undermined by it. With modern technology, human beings no longer merely live in the natural world, but they consume it and disrupt its life-sustaining processes in ever more destructive ways.

The phenomenon of technology without limits affects all who live on the planet, and it affects most directly those who live in the affluent industrial and postindustrial nations of North America, East Asia, and Western Europe. In the United States, it would be difficult to find anyone who does not have an awareness of it or does not experience it on a daily basis.

How many Americans, however, have had second thoughts about the rationality of the process of technological innovation without end that goes on around them? There may even be some who have wondered if a culture based primarily on technology and its achievements is really a culture, or if it is rather a prelude to some new form of barbarism. How many scientists and engineers have questioned the value of their own involvement in the development of technologies for which there is no real need or which threaten to bring new and serious risks to both humans and the environment? And among the prospective scientists, engineers, and technologists studying in colleges and universities, how many are concerned about the moral dilemmas and practical implications of the technology-based careers for which they are preparing?

The phenomenon of technology without limits displays, for reasons which are both practical and cultural, its fidelity to the principle of means without end in two fundamental ways. In the first place, as Jacques Ellul was one of the first to point out, the goals and norms of modern technology (the means) have become autonomous with respect to values (the end) and now govern the society as a whole. Today, however, it is not just society that is affected, but as we indicated, culture itself. The result is that we are now living in a technoculture. Second, in both theory and practice, the dynamic of development of modern technological means has no finality or point of termination. That is, the process of technological innovation is open-ended and essentially proceeds at any point regardless of what the uncertainties or negative implications of the technology in question may be. Furthermore, the question of whether humans really want it and whether they even need it is essentially ignored. In our contemporary technoculture, the process of the continual multiplication and enhancement of technological means functions for all practical purposes as an end in itself.

A primary practical reason for this phenomenon of technological innovation without end is the fact that in the prevailing context of capitalist economics, the motivating principle of all activity is the pursuit of profit for the sake of profit. Since new technologies mean new profit opportunities, they are usually welcomed by the profit-seeking corporations which dominate the economic arena. Thus, when the CEOs of the world's leading high-technology enterprises gathered in Davos, Switzerland, in January of 2001, less than a year after a meltdown in the hi-tech sector, they did not hesitate to make it clear they were as committed as ever to the technology-for-the-sake-of-technology principle. Carly Fiorina, CEO of Hewlett-Packard at the time, made a typical comment. "We're at the beginning of what technology can do, not the end," she said. "We're moving into a world where everything is going to be intelligent," she added, "and everything is going to be networked."[1]

In the same vein, Microsoft's Bill Gates predicted a proliferation of new devices, from mobile telephones and personal digital assistants to e-books and detachable PC screens. In his view, the key challenge at that time was development of software to link everything together, so individuals could communicate and access up-to-date information at the press of a button.[2] Furthermore, in an interview with a *New York Times* reporter about a new consumer product, Gates frankly acknowledged the importance for hi-tech corporations of creating a demand for new products that did not exist before their development. "This is about great software," he said, "and it's about evangelization."[3]

Paul Virilio, the French technocritic, commented on the cynical nature of such a practice, caustically referring to the "promotion of the Web and its on-line services" no longer bearing "any resemblance to the marketing of a practical technology."

"What is involved," he said, is "the most immense enterprise of opinion transformation ever attempted in peacetime," done with "scant regard for the collective intelligence or the culture of nations."[4]

Given the faithful adherence of his colleagues to the innovation-for-innovation's-sake principle, Bill Joy, a computer expert who was chief scientist at Sun Microsystems in 2000, created a stir in the technoscientific sector when he called for the termination of research that had a potential for creating a threat to the human presence on the planet. Warning of the danger in the twenty-first century that self-replicating, superintelligent machines combining nanotechnology, robotics, and genetic technology (GNR) would gain control and "replace our species,"[5] Joy stated that "The only realistic alternative . . . is to limit development of technologies which are too dangerous, by limiting our pursuit of certain kinds of knowledge."[6] He thus raised an issue that should be the subject of a prolonged debate in the scientist-technologist community, although the fact that most of its members depend upon never-ending innovation to justify research funding and to enhance their prospects for career advancement creates a powerful disincentive to any true discussion of the issue.

One of the few who did join the issue publicly was Dr. Freeman Dyson, an Emeritus Professor of Physics from Princeton who debated Joy at the Davos Forum of 2001. He provided a typical example of the self-serving response of technoscientists, however, when he based his argument on a need to protect the abstract principle of intellectual freedom. Writing in the *New York Review of Books* in February 2003, in support of his thesis, he quoted John Milton's rejection in 1644 of government-controlled licensing (i.e., censorship) of publications in his speech, *Aeropagitica*. "God," Milton said, ". . . gives us minds that can wander beyond all limit and satiety. Why should we then affect a rigor contrary to the manner of God and of Nature . . .?"[7]

To suggest that technology controls after the research stage might be acceptable, however, Dyson quoted another passage in *Aeropagitica* where Milton approved censorship of books *after* they had been shown actually to have had a pernicious effect.[8] The problem with this approach, however, is that once a lot of time and corporate or taxpayer money has been spent on research, and after those who expect to profit from the new technology have mobilized their support, even if serious questions arise about a new technology, it is very difficult to stop development and use. Furthermore, the true nature of the damage may not be apparent until *after* implementation, in which case the harmful effects may be irreversible and of an unacceptable magnitude.

He also cited the ten-month moratorium on gene-grafting research that scientists imposed on themselves in 1976 because of safety concerns[9] as an example of an acceptable form of research control, presumably because it was voluntary, temporary, and was in the hands of the scientists alone. The problem with voluntary controls, of course, is that there is no independent enforcement if there are violators; and when scientists and technologists stand to benefit financially and professionally from engagement in a given area of research, it may be unrealistic to expect they would agree to long-term controls.

Dyson also mentioned favorably the application of a risk-benefit standard to resolve questions concerning new areas of scientific and technological research.[10] This approach is often preferred by corporate representatives, government bureaucrats, and technoscientists because it makes it possible to simplify complex questions by quantifying the variables and comparing the magnitude of the anticipated risks with that of the expected benefits. A problem, however, is that the future benefits are often deliberately exaggerated (i.e., as currently may very well be the case for stem cell research). Furthermore, how can benefits which readily lend themselves to quantification (monetary profit, for example) be compared with risks so severe or disruptive that their impact will be essentially qualitative, such as aggravated warming of the earth's atmosphere or rendering a large area hostile to life for centuries by contaminating its environment with radioactive wastes from nuclear power plants? Technologists and systems analysts often find a way to assign quantitative values to such variables, but such practices are essentially inaccurate, misleading, and highly suspect.

In any case, Dyson was not sympathetic to Joy's appeal for controls on research for technologies which are perceived as capable of posing a threat to human existence. Furthermore, most of his colleagues in technoscience do not want controls on research either. And given the nature of our technoculture of technology without limits, it is likely that the vast majority of scientists and technologists will continue to pursue their investigations, moving mechanically from one discovery to the next, with no impediment other than financial constraints or the occasional concern for social and moral responsibility which may afflict a few of their colleagues on sporadic occasions.

A Focus on Ideological Developments in the History of Western Culture

Dyson's appeal to history and ideas when he cited the words of the great Puritan poet Milton, of course, was highly selective. It underscores, nevertheless, the importance of cultural factors and the need for a historical perspective to understand the origins of our present technoculture of technology without limits and to evaluate the validity of its basic principles and practices. To merely study the history of the progress of technology per se, important as it is, without taking into account the broader context of ideological developments, is not enough. Not only should the examination of ideological transformations over the centuries help to explain why higher values no longer control technology in our postmodern technoculture, but it should also provide greater insight into the emergence of the phenomenon of technological innovation without end.

This book contains, therefore, a historical overview of broad transitions in philosophy, theology, and science relevant to our inquiry, as well as a survey of ideas concerning technology itself. Some of the latter ideas guided technological development and use, some encouraged or discouraged technology as a human activity, some provided legitimization for technology and its broader effects, and others, starting with the industrial era, made a critical analysis of technology and its impact on life and culture.

Although other civilizations—China, for example—played a significant role in the development of technology up to the time of the Scientific Revolution in Europe, we believe the roots of contemporary technoculture and the phenomenon of technology without limits lie primarily in the West. We concentrate, therefore, on technology-related ideological developments in the history of Western civilization, from the classical Greeks, who considered *techne* to be neutral means subject to higher values, to the postmodern present, in which contemporary philosophers tell us durable higher values are missing and where technology theorists see information technology as providing the goals and norms that inform, for better or worse, culture and society as a whole. The book provides an overview, in essence, of some of the most important ideological elements in the history of the West's evolution from a civilization of ends to a civilization of means in less than 3,000 years.

For the purpose of the analysis, we have divided this time frame into six historical segments. The first is the Classical period of Apollonian Greek culture, when everything, including *techne*, was characterized by a concern for limits and subject to guidance by higher values. Greek philosophers emphasized living according to nature, rather than mastering it by means of technology.

Second are the Patristic and Medieval periods of Christian civilization, when theologians linked technological development to the realization of human destiny and laid an ideological foundation which served in the long run to undermine the idea that technology should be subject to limits. These ideological structures included such themes as the anthropocentric nature ethic of Genesis, the goal of human recovery from the Fall, the idea of human likeness to God, the finishing/furnishing-of-creation theme, and a teleological view of history, which eventually included the idea of a millennium. Technology made impressive advances during the Middle Ages, although science, with the exception of alchemy, remained largely useless for practical purposes, immobilized as it was by the strictures of Christian theology and by a later revival of teleological Aristotelian nature philosophy.

The third period encompasses the Renaissance and Scientific Revolution, during which time the influence of alchemy and the rediscovery of mechanistic Greek nature philosophy led to the new, technologically applicable power knowledge of Francis Bacon and René Descartes in science and the formulation of a collective human project to master nature. Henceforth, scientific discovery and technological innovation were closely linked.

The fourth period is the Enlightenment, when the Christian theodicy of inevitable benefit from historical change was applied to technological innovation in a new, secular theory of progress and when existing ideas, including those pertaining to God, religion, history, government, knowledge, and reason itself, were subjected to critique by the *philosophes*. In opposition to other leading thinkers of the time, however, Jean-Jacques Rousseau articulated his famous rejection of the theory of progress and qualified technology as a source of human misfortune, not benefit.

The fifth period is the Nineteenth Century, when the Industrial Revolution gained its full momentum. Karl Marx and Friedrich Engels were the philosophers who dominated the ideological landscape in technology theory. They argued that its level of development determined the level and character of ideas, life, and culture. In making this assertion, they gave the theory of progress a critical and revolutionary dimension within the existing context of industrial capitalism, although they notably failed to apply their critical approach to technology per se.

In the century's closing decades, competing paths to truth and fulfillment based on religion, reason, history, science, and art were rejected by Nietzsche, who proclaimed the advent of nihilism and the rejection of all higher values. Western culture, in other words, had entered into a stage of crisis, and philosophy seemed to have reached an impasse. Disappointment with the results of democracy, capitalism, and industrialization led some late nineteenth-century artists and writers to advocate a new *techne* of art, morally validating itself by its beautiful style rather than its content. In making this claim, artists and writers of the Realist school prepared the way for the morally autonomous theories of technology of the reactionary modernist and National Socialist thinkers of the early decades of the twentieth century.

It was in the sixth period, the Twentieth Century, that the phenomenon of technology without limits emerged and eventually came to define a new culture, a technoculture. In the early decades of the century, reactionary modernist philosophers and National Socialist ideologues, clearly influenced by Friedrich Nietzsche, emphasized a will to power; but unlike Nietzsche and G. W. F. Hegel, they associated technology with what they considered to be the positive, Germanic values of *Kultur*. They reified technology and treated it as a thing in itself, linking it with German will and soul. And despite its dangers for human freedom and autonomy, they glorified technology and embraced it in their new totalitarian state. From the conservative right, Martin Heidegger dissented, warning that technology threatened the "disclosure of being" and suggesting that art might be the only salvation from the "enframing" nature of technology.

The emergence at mid-twentieth century of advanced industrial society, in which technology's own norms and methods infused most aspects of life and culture, led to Ellul's critical analysis of what he called a *technological society*, precursor of our present technoculture. Ellul's thinking about the importance and effects of technology therefore went beyond Marx and Engels, who had said only that the level of technology determined the character of life and culture of the society in which it existed. Herbert Marcuse, who was a contemporary of Ellul, demonstrated how logical positivism and other philosophies that concentrated on the analysis of language had lost their critical role, allowing the existing social order to escape any effective ideological challenge and to legitimize itself by its technological achievements alone.

With the emergence of a nihilistic postmodern philosophy in the 1970s, typified by Jacques Derrida's deconstructionism, and with the impact of information technology in the 1990s, technoculture finally appeared on the horizon. Jean-François Lyotard argued that absent any credible philosophical "grand narratives," the legitimizing principle for postmodern society should come from the "language game" of cutting-edge scientific investigation, paralogy, intended as a way to discover the new. He assigned a crucial role in this search to computerized data banks. Neil Postman, however, made a critique of what he termed *Technopoly*, our technocentric culture in which information technology currently plays a dominant role. He argued that contemporary life is afflicted by a glut of information which, like the impact of AIDS on the human body, has overwhelmed all existing filters and produced a kind of cultural anomie or pathology.

Jeremy Rifkin explained how biotechnology, perhaps the most profound in its implications of anything in the history of technology, is, practically and theoretically, an extension of the principles of information science into the realm of biology. That is, a new paradigm for organic life and a new, biotechnology-friendly theory of evolution have been derived from cybernetics. Armed with these ideas and the ability to alter the genetics of living organisms and to create, even, transgenic organisms, genetic scientists and biotechnicians have embarked on a new adventure in technoscience without limits, not without its ethical, genetic, and ecological contradictions.

Finally, at the end of the second Christian millennium, Virilio's controversial writings represented a darker and more pessimistic strain in the technology critique. That is, he argued that the digital technologies of the Information Age have created a new danger by eroding normal cognitive and sensory links of humans with the real world of time and space in which they live. He warned of an apocalyptic potential for a "total accident," given the destructive power and scope of what he called the "three bombs"—information, genetic, and nuclear technology. According to Virilio's fatalistic view, the nature of these technologies and their advanced level of development mean it is already too late to restore ethical limits or establish reliable practical controls.

Notes to Introduction

1. *International Herald Tribune*, 30 Jan 2001.
2. Ibid.
3. *New York Times*, 5 Jan. 2001.
4. Paul Virilio, *The Information Bomb* (New York: Verso, 2000), 110.
5. Bill Joy, "Why the Future Doesn't Need Us," *Wired* 8.04 (April 2000).
6. Ibid., 12.
7. Freeman Dyson, "The Future Needs Us," *New York Review of Books,* 3 Feb. 2003: 13.
8. Ibid.
9. Ibid., 12.
10. Ibid., 11–12.

1

The Classical Greek Period

Apollonian Greek Culture
vis-à-vis the Superior Power of Nature

NATURAL PHILOSOPHY: THE APOLLONIAN NATURE PARADIGMS

In the view of Camille Paglia, the origins of Western science can be seen as the product of the Apollonian strain in classical Greek culture, representing an effort of human Logos to counteract the fearful power, mystery, and cruelty of nature by means of naming and classifying, and by providing a plausible, logical analysis of its operations.[1] She was careful to add, however, that what the Apollonian Greeks sought to overcome in nature by classifying and explaining was in its bowels, the dangerous "chthonian" element, rather than in its more benign surface.[2] This included the "blind grinding of subterranean force . . . the dehumanizing brutality of biology and geology . . . Darwinian waste and bloodshed . . . squalor and rot . . ."[3]

There were two basic Apollonian explanations of nature that had the most significant influence on Western science during the last 2,500 years. One was a mechanistic view, which portrayed nature as a mechanical entity without higher purpose. It had its most enduring expression in the form of the atomist paradigm, represented by such philosophers as Democritus and Epicurus. The other was the teleological paradigm of thinkers like Plato and Aristotle, who believed nature had a final cause, or higher purpose.

Not only in classical antiquity, but again during the Scientific Revolution of the sixteenth and seventeenth centuries, the teleological and the mechanistic viewpoints vied for preeminence. The teleological paradigm, with its idea of a purposeful world, was compatible with Christian theology and had been integrated into it by means of Aquinas's Scholastic synthesis of the thirteenth century. When the mechanistic atomist paradigm of Democritus and Epicurus reappeared on the cultural horizon three centuries later, however, it was an important catalyst for the construction of modern science, despite the resistance from religious authorities generated by its prima facie incompatibility with Christianity.

The mechanistic atomist paradigm of nature. There were a number of Greek pre-Socratic nature philosophers whose theories, based on reason alone, influenced scientific thinkers centuries later. In the sixth century B.C., for example, Anaxamander theorized that an infinite and eternal primal substance had separated into various forms with opposing properties such as hot and cold and wet and dry. The earth, he said, was initially a watery mass surrounded by fire which eventually dried out. He added that the first form of life crawled out of the warm slime; and, anticipating the modern theory of evolution, he stated that humans, like all animals, had descended from fish.[4]

Heraclitus, a mechanistic nature philosopher of the fifth century B.C., declared that the basic characteristic of nature was perpetual change involving a struggle of opposites. His belief that fire was the primary substance prefigured Einstein's theory that all matter was energy, according to Heisenberg, the famous twentieth-century physicist.[5]

Empedocles, who died in the fourth century B.C., came up with the influential theory of the four elements: air, water, earth, and fire. It resurfaced in the Renaissance with the addition, by Aristotle, of a fifth celestial element. Anaxagoras, a contemporary, theorized that an infinite variety of infinitely small "seeds" of matter had configured into the things of the world as the result of a rotary process produced by Mind.

Over the long run, however, the most influential of these nature philosophers were the atomists. Democritus, in the sixth century B.C., and Leucippus, of whom little is known, in the fifth century, were the originators of the atomist paradigm of nature. Epicurus (fourth century B.C.) based much of his nature philosophy on Democritus; and Lucretius, a Roman of the first century B.C., turned Epicurus's philosophy into poetry in his *De rerum natura.* Centuries later, in the Renaissance, his work was an important vehicle for reintroducing atomism into European thought.

The atomists saw nature as a self-operating mechanism, devoid of any higher purpose and made up of tiny particles of matter called atoms, which had come together in various combinations. Democritus believed that everything was subject to blind fate and that it was impossible to determine the cause of the atoms, which, like entire worlds, were of infinite number. Epicurus also believed in an infinity of time and worlds, and believed the universe, composed of atomic particles, was ruled by chance. In the face of this absence of purpose in the world, Democritus and Epicurus both formulated their responses to nature primarily in defensive terms. For them, the answer was *ataxaria*, or cultivation by the individual of a life without care or disturbance, based on their philosophy.[6]

The teleological nature paradigm. The teleological conception, according to which everything in the natural world had not only a rational explanation but also a higher purpose, was more reassuring to humans than the view of the atomists and the other pre-Socratic nature philosophers we have mentioned. For example, in the *Timaeus* dialogue, written in the fourth century B.C., Plato expressed the view that all living things were part of a chain of being wherein everything had a predetermined rank and purpose and where humans occupied the highest position.[7]

Aristotle shared this perspective, as revealed by his description in *De Anima* of nature as a hierarchical order in which every living thing had its own particular purpose or end (*telos*).[8] His idea that God was the object of all desire— and the final cause of all activity of living things—was another indication of his belief in teleology.[9] Furthermore, both he and Plato described the universe as an orderly, hierarchical, and finite system, with earth at the center and the planets and fixed stars rotating in circular paths around it.[10]

In the Apollonian sense of providing a logical account of nature, the teleological paradigm was a more fully realized explanation of nature than atomism and the other mechanistic paradigms. Not surprisingly, its adherents framed their response to nature in more positive and ambitious terms. It was based on the philosopher's attainment of knowledge of the good and the higher form of happiness that would result for humans if they lived a life guided by it. Its most illustrious representatives were the philosophers Socrates, Plato, and Aristotle, archetypal examples of the Apollonian sky cult that left its mark on high classical Greek culture.[11]

Because of this very concept of purpose, however, the teleological conception of nature provided a strong ideological disincentive for human attempts to tamper with nature and its processes by means of *techne*. That is, what right or need did humans have to intervene in nature's operations if everything in it belonged to a preordained order and had been assigned a particular purpose? The mere existence of such an order suggested that there were limits in nature which had to be respected.

PHILOSOPHY AS THE ANSWER TO THE SUPERIOR POWER OF NATURE

Neither the teleological conception nor the mechanistic approach encouraged humans to marshal a collective effort to alter the character of their relation to nature. Nor did either provide any theoretical knowledge which could have been applied technologically for the practical mastery of nature. Both, however, were incorporated in a broader philosophical context which offered individuals a way to work out their existence in the face of the superior power of nature. Each approach included the basic Apollonian principles of limit and just proportion, mind over nature, and the idea that life guided by philosophy would be rewarding. And as we have indicated, mastery of self, guided by philosophy—not mastery of external nature by means of *techne*—was the Apollonian answer.

The Apollonian Principle of Limits

HUMAN NATURE TO BE GUIDED BY THE PRINCIPLES OF SKY-NATURE

Although the Apollonian Greeks' response to nature was not one of mastery and control, their approach to human nature, which was subject to control by means of social and ethical mechanisms, was exactly the opposite. They wanted humans, in effect, to subject their lives to the control of higher cultural principles derived from the order, regularity, and beauty of heavenly nature or from a principle of balance or harmony inherent in the cosmos itself. In other words, what belonged to earth—nature, in the case of human beings—had to be subjected to the higher principles inherent in sky-nature or the cosmos itself. This is what the historian Michael Grant meant when he said living by nature's principles was the very essence of classical Greek culture.[12]

Mechanistic nature philosophers such as Empedocles, Heraclitus, and Anaxamander all identified a principle of "justice" or balance in the cosmos itself. For Empedocles, it was in the process by which the cosmic forces of Love and Strife acted to combine or separate the elements, each prevailing at a given point but neither succeeding permanently. For Heraclitus, cosmic justice was assured by a never-ending process of contending opposites that prevented a complete victory by either. Anaxamander had a similar view, according to which elements like fire, water, and air were constantly seeking to enlarge their empire in the world but were blocked from doing so by a kind of natural law that restored balance.[13] According to each of these theories, therefore, there was an inherent dynamic that acted to place limits on the action of the primary cosmic forces.

Plato's teacher in real life, and the central figure of his *Dialogues*, Socrates, stated in the *Republic* that the sky was "the loveliest and most perfect of material things" and had been "framed by its artificer with the highest perfection of which such works are capable."[14] For this reason, he added, the study of astronomy was "a means to the knowledge of beauty and goodness."[15] It revealed, in effect, a nature of order, regularity, beauty, and just proportion. These were the natural—and rational—principles that were incorporated into the fundamental Apollonian cultural formula for right living and happiness—the respect for limits.

When Socrates asserted that each individual in his ideal society of the *Republic* should occupy the role and station for which he was best suited by nature, for example, he made it clear that this did not mean behaving according to one's natural instincts.[16] On the contrary, he emphasized the importance of temperance as a basis for happiness, both in the individual, where the rational element would have to rule the lower, bodily elements, and in society as a whole, where the "wise" segment of society—philosophers—would have to rule the more numerous and base element.[17]

The key ethical debate in Plato's *Republic* was the argument between Socrates and his Sophist adversary, Thrasymachus, over the broad question of limits. In that debate, Socrates succeeded in getting Thrasymachus to admit that a wise person was one who knew how to stay within limits.[18] The worst—and most unhappy—person, according to Socrates, was the political tyrant, whose aberrant power enabled him to indulge his every whim and violate all reasonable limits.[19]

The Olympian pantheon of deities included, as a matter of fact, Nemesis, a goddess whose precise role was to punish those individuals who were guilty of hubris and violation of the principle of limits in their actions. Carrying a measuring rod, a sword, a whip, and a bridle, and riding in a chariot drawn by griffins, she avenged evil and punished arrogance. Her just retribution thus served to preserve the rhythm of fate and to maintain the essential equilibrium of the human condition.[20]

ARISTOTLE'S SUBORDINATION OF PRODUCTIVE KNOWLEDGE TO PHILOSOPHY

True to the Apollonian ideal, Aristotle, too, believed that the avoidance of excess and the respecting of limits was the cardinal principle of moral truth. In his *Nicomachean Ethics*, he stated that "virtue is a kind of mean . . . determined by a rational principle." "Evil," he said, "belongs to the state of the unlimited."[21]

Aristotle applied this principle of limits to knowledge itself by establishing a hierarchy of knowledge according to which the lower categories of truth were subject to control by the higher ones. He distinguished, for example, between productive knowledge, practical knowledge, and philosophical knowledge. Productive knowledge, which included *techne*, could never be done for its own sake and had to serve an externally supplied end, or *telos*. Practical wisdom, he said, was concerned with a reasoned state of capacity for action (i.e., *praxis*), which included politics and ethics.[22]

Philosophy, in contrast to the other branches of knowledge, investigated the underlying causes and the fundamental principles of things. Reason, Aristotle said, was the divine element in humans. And the highest form of human activity, pursued for its own sake and thus the source of the greatest happiness, was to devote one's reason to the contemplation of philosophical truth.[23] When humans engaged in rational contemplation, he added, they were imitating God, who was self-thinking thought, removed from the world and located in the heavens.[24]

LIMITS ON HUMAN ACHIEVEMENT IMPLICIT IN CYCLICAL GREEK HISTORY

The Apollonian principle of limits was also implicit in the Greek belief that, in contrast to the linear and teleological Judeo-Christian view, history was cyclic. In other words, there was a limit to what humans might accomplish before events reverted to misfortune or failure. Aristotle, for example, stated that events and opinions, "as they occur among men, evolve not only once or a few times, but indefinitely often."[25] And Thucydides, the great Greek historian of the fifth century B.C. also referred to a cyclic pattern of recurrence in human events when he wrote: "[R]evolution brought upon the cities of Hellas many terrible calamities, such as have been and always will be while human nature remains the same . . ."[26]

According to Werner Jaeger, the belief that there were many contending and often capricious gods, rather than a single God who directed human affairs, was one of the primary reasons why the Greeks did not believe that history had a consistent direction toward a higher goal.[27] Nor did they believe that humans had the capacity to shape freely their destinies on their own, either individually or collectively. They adhered, rather, to the idea of inalterable—and often tragic or absurd—individual fate (*moira*), which placed an arbitrary limit to what anyone might accomplish or experience.[28]

The Apollonian *Techne* of Limits

The virgin goddess Athena, guardian of the Athenians and a dominating presence in Homer's *Odyssey*, represented the mind in Greek literature and mythology as *techne*, according to Paglia. She was divorced from nature and symbolized the human-made; and for the Athenians, she was the patroness of the crafts and the cultivated olive.[29] Her intelligence was "contriving," in

Paglia's words, and able to "subdue circumstance to will and desire." Thus, Paglia stated, Athena favored men of action like Odysseus, a "man of many wiles." Her power, she concluded, symbolized the "fanatical purposiveness of the West, limited but all-achieving."[30]

The Greek definition of *techne*, or "art," included any rational method or procedure designed to produce a predetermined result. It applied, for example, to a professional or vocational skill that produced a particular benefit or effect on humans who were its object; or it could have been a craft or process that produced a tangible, material thing. To be wily and clever in the manipulation of nature, in any case, was the essence of *techne*. The possession of philosophy and wisdom, however, was no more necessary for successful mastery of the skills of a *techne* in ancient Greece than it is today, although Plato and Aristotle argued that respect for higher values was necessary for its proper use.

THE PROMETHEUS MYTH'S TEACHINGS ABOUT *TECHNE*

The most important Greek myth about *techne* was the myth of Prometheus. It not only influenced later thinkers who wrote about technology, but it contains important and enduring truths about technology. According to the story, Prometheus and Epimetheus were primeval gods called Titans, who assisted in the creation of living creatures in the world. Epimetheus, whose name means "the one who thinks afterwards," unwisely distributed all of the characteristics he had to assign to the animals alone, which were devoid of reason. They thus had claws, strength, fur, tough hides, horns, wings, speed, courage, etc. On the other hand, humans were left with only their upright stature and power of reason. Otherwise, they were naked and unarmed.

At this point Prometheus, whose name means "one with foresight" and who was to oversee this distribution process, decided to steal the secrets of *techne* from the gods and give them to humans so they could ensure their self-preservation and survival. From Hephaestus (Vulcan), the son of Zeus, he stole the techniques for using fire; from Athena he stole the secret of the other *technes*. Humans thus acquired the knowledge and skills necessary to engage in agriculture, build and heat shelters, make clothing, fashion weapons and tools, and engage in other technical pursuits.

The myth relates that Zeus, the supreme god, was angered by Prometheus's audacity and decided to punish him as well as the humans who had received the secrets of *techne* from him. Accordingly, Zeus ordered subordinate deities to create a perfect woman, Pandora, whom he sent to earth bearing a closed box containing all sorts of evils and crimes. Prometheus, who possessed foresight, was suspicious and avoided her. But his brother, Epimetheus, despite warnings, imprudently took Pandora as his wife. He allowed her to satisfy her curiosity and open the box, at which point all the evils of the universe, including spite, envy, rheumatism, gout, and so on, spilled out. All that remained in the box was hope.

Zeus ordered a different punishment for Prometheus. He commanded Hermes (Mercury), according to one version of the story, and Hephaestus, according to another, to chain Prometheus to a rock on Mount Olympus. Every day a vulture came and picked at Prometheus's liver, which continually regenerated itself. Prometheus's suffering was terrible, however, until Zeus finally allowed Hercules to release him.

Thus, like the relation of the events of the Fall in Genesis, the Prometheus myth tells a story of transgression, human acquisition of forbidden knowledge, woman as the agent who brought lasting misfortune to humans, and divine punishment visited on all of humanity. It also contains a number of basic conceptions associated with technology.

First, there is the idea that technology represents a human usurping of godlike powers. Second, the myth includes the idea that technology involves a form of rebellion against the original Creation. Third, it promotes the belief, re-iterated in the first century B.C. by Cicero in *De natura deorum*, that technology is a basis for human uniqueness in relation to animals. And fourth, the myth contains the idea that there is a price to pay for possession of technological power over nature. Consistent with the punitive consequences for humanity described in the Pandora story, we pay this price today in the form of noxious pollution, dehumanization, resource depletion, the existence of weapons of mass destruction, and other technology-generated problems and dangers.

Socrates, who related part of the story of Prometheus in the *Protagoras* dialogue, showed his sensitivity to the fact that *techne* alone was not sufficient for a civilized existence. He explained that the humans who acquired *techne* from Prometheus lacked the knowledge of political skills, and that this omission led to injustice, wars, and massacres. That was why, he said, Zeus dispatched Hermes to bring to humans a sense of law and justice so they could live together in greater harmony.[31]

THE ICARUS MYTH'S WARNING ABOUT TECHNOLOGICAL HUBRIS

Another important Greek myth relating to *techne* was the story of Dedalus and Icarus. Dedalus, an engineer, built the famous labyrinth in the palace of King Minos of Crete, who had confined him to the island after he made a device which facilitated the consummation of his queen's illicit love for a white bull that had emerged from the sea. To escape from Crete, Dedalus fashioned wings from feathers and wax for himself and his fifteen-year-old son, Icarus. He warned his son to fly neither too low nor too high because the sea's humidity or the sun's heat could damage the wings. Once airborne, however, Icarus was exhilarated by his new powers of flight. He rose higher and higher, closer and closer to the sun. The wax that held the feathers of his wings together melted, and he plummeted to his death into the sea. Dedalus, however, continued his flight and, according to different versions of the story, ended up in either Sicily, Italy, or Egypt.[32]

This myth is a good illustration of the Apollonian concern for the respecting of boundaries and limits, applied in this case to a *techne*. On the other hand, the myth could also be read as illustrating what can happen to humans if they become too Apollonian, too "sky-bound" (in both a figurative and a literal sense, in this instance) and lose the feeling of connectedness with nature. The story of Icarus reveals, too, the danger for humans who are so infatuated with newly acquired, technological power that they lose their capacity for reasoned judgment. It also warns that dependency on technology alone can be very unwise, for if it fails, everything may fail.

SOCRATES' LINKAGE OF *TECHNE* AND SOCIAL RESPONSIBILITY

In their famous discussion of justice in Plato's *Republic*, Socrates and Thrasymachus examined the questions of what the proper goal of a *techne* was and whether its application should be subject to any limitation. Thrasymachus took a subjective approach and maintained that one practiced a *techne* only for one's own benefit. Socrates argued that true knowledge had to be objectively valid and that a *techne* was always properly done for the good of others. He cited the example of the doctor, who used his *techne* of healing for the benefit of his patients.[33] This position reflected the Apollonian concern for limits. To put it in contemporary terms, we could say that the practitioner of a *techne* had to be socially responsible.

ARISTOTLE'S DEFINITION OF *TECHNE* AS NEUTRAL MEANS

The fifth century B.C. atomist philosopher Democritus saw craftsmanship as a kind of manual dexterity, close to instinctive animal capabilities. He characterized architecture and weaving, therefore, as similar to the building of nests by birds and webs by spiders.[34] In contrast to this emphasis on native ability, the Stoic philosopher Posidonius (second and first centuries B.C.) argued that the idea of the rudder came from conscious imitation of a fish's tail.[35] Aristotle, too, saw *techne* as *mimesis*—a deliberate imitation of nature—although for him, the imitation was not in what was produced but rather in the process by which the maker gave form to it.[36]

According to Aristotle, all things that existed in the world either came to be by nature, whose principle of movement was immanent, or were produced by art (*techne*) which required an external source of movement.[37] *Techne*, he added, involved a reasoned capacity to make and produce something (not necessarily a physical object) that did not previously exist.[38] In either case, natural or arti-ficial, the process that determined the product or result could be understood in terms of the four causes—material, formal, efficient, and final. In his view, they governed all change in the world of existent beings.[39]

For example, when the craftsman used *techne* and produced a thing, he (the efficient cause) imitated nature in uniting form (the formal cause) and matter (the material cause). The essentially Apollonian element of Aristotle's theory of *techne*, however, was the requirement that it serve an externally supplied end, a "that for the sake of which," or final cause (*telos*). *Techne*, therefore, was only, in Aristotle's words, "means toward the end."[40] As such, it belonged to the inferior category of productive knowledge, which was never done for its own sake. In the case of a making technique, the *telos* was the purpose for which the thing was made; and in the case of a *techne* that used the product, it was the purpose for which it was used.[41] In both instances, the purpose was informed by external, higher values. The necessary element of final cause, therefore, meant that Aristotle's theory of *techne* was one of limits.

Writing in the first century A.D., the Roman natural historian Pliny seemed to echo Aristotle when he stated that a specific *techne* could be used for both good or bad ends, its morality being a matter of human application. "It is by the aid of iron" (i.e., a product of mining and metallurgy techniques), he wrote, "that we construct houses, cleave rocks, and perform so many other useful offices of life. But it is with iron also that wars, murders, and robberies are effected."[42]

This Aristotelian idea of *techne* as neutral means is still popular today, often invoked by those, ironically, who do *not* want any limitations on technology at the crucial stages of research and development. They argue that since tech-nology, according to the Aristotelian theory, is neither good nor bad, harmful ef-fects are always only a function of the choice of use, rather than the inherent characteristics of the technology itself.

According to Aristotle, philosophy included the sciences, which became possible in time only after making (*techne*) had provided a material infra-structure which permitted leisure. To illustrate this point, he explained that mathematics were founded in Egypt *after* material conditions were sufficiently advanced to provide the necessary leisure for a priestly caste to develop them.[43] Thus, *techne* made it possible for humans to engage in intellectual pursuits and have a cultural life of a higher level.

For Aristotle, however, it was philosophy that provided the guidance as to how this life should be lived. Although philosophy was not prior in order of invention, he recognized it was of a higher order of importance. It is thus clear from Aristotle's thought that measuring the quality of human existence in terms of the level of technological prowess alone does not accurately reflect the cul-tural level of a society or civilization. It is inconceivable that he would have approved of anything like our present technoculture, in which technology at the most advanced level has absorbed higher culture and threatens to trivialize human existence in a civilization of means.

Aristotle differed from Plato by including poetry in the category of *techne*.[44] His inclusion of this and other fine arts in the classical *techne* was important because in the twentieth century, reactionary modernist and Nazi ideologues deliberately merged the two *techne*s, art and technology, by aestheticizing technology. They argued that because it was beautiful, technology, like the Realist art of the nineteenth century, was an end in itself. After the defeat of Nazi Germany, however, Heidegger suggested that the *techne* of art might provide a way to overcome the cultural crisis of a world threatened by too much technology.

Obstacles and Disincentives to the Development of *Techne* in the Thought of Plato and Aristotle

PLATO'S DISDAIN FOR THE PHENOMENAL WORLD OF THE SENSES

Besides his teleological perspective, there was another reason why a key Apollonian thinker like Plato was not interested in the practical mastery of nature: He downgraded existence in the phenomenal world, which he considered, even if it was not devoid of purpose, as ontologically inferior to a transcendental, nonmaterial ideal world. This prejudice against the sensory world not only was a disincentive for scientific thought in Plato's own time, but it persisted as such in the Neoplatonist philosophy of Plotinus in the third century A.D. and in the otherworldly focus of Augustine and other medieval Christian thinkers.

According to Plato, true reality, true Being, and the Good, were located in the ideal world. It was there that the Forms, abstract ideas that were the prototype essences of the imperfect, particular things of the sensory world of phenomena, actually existed. Plato had a greater interest in trying to know these Forms by means of philosophy, particularly the Form of the Good, than in seeking to understand the physical workings of the imperfect world of nature. Knowledge of the Good, he believed, provided the basis for happiness.

The main exceptions to Plato's general disregard for scientific questions were some superficial points of cosmography in the *Republic*'s Myth of Er; his theory of the origin of the world and description of a hierarchical chain of created, natural beings in the *Timaeus*; and his promotion in both of these dialogues of the study of mathematics and the heavens. These excursions into what today would be called science, however, remained only on a level of

Apollonian explanation and provided no knowledge that could be practically applied for the manipulation and control of nature.

Plato thus bequeathed to European philosophy an Apollonian prejudice against the reliability of the senses for true knowledge—that is, a tradition of negative thinking based on the belief that reason had to penetrate the deceptive sensory facade of things in the phenomenal world to discover a deeper reality. Such a prejudice, however, valuable as it might be for critical philosophy, downgraded the very empirical knowledge that Bacon, in the seventeenth century, said was necessary in science to provide the basis for human power over nature.

In Plato's view, the mind had to detach itself from the sensory perceptions of the lowly things of the earth if it were to attain the higher truth of the Forms, including the idea of the Good upon which a perfected life had to be based. In the *Republic*, Socrates described the philosopher's difficult quest for philosophical knowledge by means of his famous Allegory of the Cave. It started with a prisoner in a dark cave, where his sensory perceptions of shadows on a wall, which he took to be reality, symbolized the lowest level of knowledge, the opinion and ignorance of the many. The allegory ended when he escaped, emerged above ground into the light, and finally was able to look at the sun itself, representing the Form of the Good. This act typified the superiority for Plato of rational, philosophical truth over sensory knowledge of phenomena.

The one important exception to Plato's disdain for empirical knowledge was the observation of the higher, more perfect sky-nature. It had value in his eyes primarily for uplifting the mind and preparing it for philosophy, not for what we would consider today as scientific purposes. As related by Plato in the *Timaeus,*

> The motions, which are naturally akin to the divine principle within us are the thoughts and revolutions of the universe. These, each man should follow, and correct the courses of the head which were corrupted at our birth by learning the harmonics and revolutions of the universe . . .[45]

THE USELESSNESS OF ARISTOTLE'S SCIENCE FOR THE DEVELOPMENT OF *TECHNE*

Reassurance of order and purpose. Unlike Plato, Aristotle did not downgrade the ontological importance of things that existed in the sensory world of phenomena. According to his natural philosophy, the form of each material thing in the phenomenal world was immanent in it and enabled it to serve its ultimate purpose, or *telos*. The phenomenal world of nature was for Aristotle, therefore, the locus of true being. He did have a significant interest in explaining the workings of nature, therefore, and his philosophy did have a physics, as well as writings on zoology and biology. This difference explains why, with the recovery of Greek learning in the Renaissance, Plato's main influence was in the domain of philosophy (Neoplatonism), whereas Aristotle's physics provided the theoretical basis for European science until it was rejected and replaced by a new power knowledge in the sixteenth and seventeenth centuries.

Aristotle's explanation of the four causal principles—material, formal, efficient, and final—that governed things that "come to be by nature" contained the essential elements of his theory of nature. Its purpose, however, was not to provide practical knowledge that humans could use to master nature, but to provide humans with a reassuring, Apollonian explanation of it as a rational and purposeful entity. This aspect of Aristotle's natural philosophy was an example of what Francis Bacon later derisively characterized as the "idols of the tribe"— that is, a tendency to explain everything in terms of human needs instead of its own truth. The problem was that it did not provide any theoretical knowledge that could be practically applied to manipulate or control nature.

For example, to explain the phenomenon of motion, crucial in a phenomenal world where everything was engaged in a process of becoming, Aristotle said that all natural motion had an internal source, the inherent process by which the potentiality of the matter of a thing moved to actualize its form. That is, all natural things (plants, animals, their parts, and the four elements—air, water, earth, and fire) in the terrestrial realm of the universe had an inherent tendency, unless they encountered a hindrance, to move to actualize their predetermined form.[46]

For the purpose of a scientifically valid physics, this theory had the insurmountable disadvantages of excluding all relativity of motion and lacking any applicability if matter were redefined as inert and only moved by external forces.[47] Another difficulty was that Aristotle gave no satisfactory theoretical explanation for what he considered "unnatural motion," such as the raising of an object by means of a lever. He termed this kind of motion "violent" or "unusual," the result of action by an external source.[48]

To explain why heavy objects fell from heights to the ground, a terrestrial motion that he considered to be natural, Aristotle said they had a tendency to seek their natural place, the center of the earth, which was the center of the center of the cosmos.[49] On the other hand, he considered the motion of projectiles fired in the earth's atmosphere to be "violent" and therefore dependent on an external mover. This was why the Aristotelians of the late Middle Ages erroneously believed that such motion was rectilinear and that the moment a mover— in this case unseen—ceased to be in contact with the projectile, it would fall straight to the ground.[50]

Since Aristotle maintained that all things above the moon were made of a fifth element, or *quintessence*, which was not subject to generation and decay like terrestrial objects, he could not explain the motion of heavenly bodies in terms of an inherent tendency to go from potentiality to actuality. He adhered to the convenient view, rather, that celestial motion was eternal, circular, and had been initiated by an unmoved mover.[51] When the revolution in science took place during the sixteenth and seventeenth centuries in Europe, this erroneous theory, along with his erroneous explanations of the motion of terrestrial bodies, was rejected.

***Epistemology without practical value for* techne.** In contrast to Plato, Aristotle accepted the reliability of sensory knowledge. By means of rational analysis of what was observed, he said, humans could discover the self-evident first principles of knowledge. That is, he believed that if one observed a thing in nature during its entire period of development and decay, its form (i.e., mature configuration) and final purpose—that is, *telos*, "that for the sake of which," could be determined. His primary interest, however, was in fitting everything into his broader philosophical system (a shortcoming that Francis Bacon scornfully identified in the Renaissance as typifying the "idols of the theater") and, as we have indicated, in providing an explanation that was reassuring in terms of human needs and purposes.

Once he had established the most fundamental truths and matters of empirical fact, however, Aristotle relied on logical inference by means of the syllogism as a tool for expanding knowledge. Like Socrates and Plato, he believed the truths he had discovered were objectively valid. As Bertrand Russell observed, he introduced the idea of category into logic, but what he defined as universals were not essential truths but only adjectives or nouns that served as class names in linguistic statements.[52]

Aristotle's approach to knowledge, while well suited to the creation of a philosophical system that would satisfy the Apollonian need to explain the totality of reality in rational and reassuring terms, produced nothing more than a static explanation of phenomenal reality. That is, it could not be expanded or corrected because of the a priori character of its basic premises and its emphasis on logic, not induction, as a primary epistemological tool. This was one of the reasons why, when Aristotle was rediscovered by Europeans in the thirteenth century, a Christian thinker like St. Thomas Aquinas was able to make a workable synthesis of his natural philosophy and Christian dogma. It was also why Bacon, who wanted a new scientific knowledge that would yield technological power over nature, subsequently called for smashing the four "idols of the mind," meaning the received truths of Aquinas's Scholastic philosophy.

The shortcomings of the epistemological approaches of Plato and Aristotle meant that technological development in the Greece of their times would essentially proceed independently of science. They did not mean that there would be no *techne*, however, but rather that it would be closely identified with the individuals who used it and that it would not reach a level of development such that it would be considered as an independent phenomenon, threatening to escape control by higher values or radically alter the relation between humans and nature.

DISPARAGEMENT OF THE MANUAL TRADES

In Plato's *Timaeus* dialogue, God the Creator was portrayed as a kind of artisan or technician—that is, as one who possessed *techne* and formed the cosmos and everything in it out of preexisting matter. In the *Republic*, Socrates used the artisan as a model for the true statesman.[53] In the *Ion* dialogue, he said painting and sculpture were *technes*. He excluded poetry, however, on the basis that it derived from divine inspiration and "Bacchic transport" and did not have the fixed rules and rational basis which characterized a valid *techne*.[54] And generally, he described the manual crafts as a vulgar form of *techne* that debased the body and the mind. In the *Republic*, for example, Socrates pointedly insisted that everyone had agreed that "the manual crafts . . . were all rather degrading."[55]

The contempt Socrates frequently expressed for the practice of manual techniques was consistent with Plato's ontological downgrading of the phenomenal world and the emphasis in Apollonian culture on living one's life according to reason rather than on the level of material conditions of existence. For both Plato and Aristotle, contemplation, not praxis, was the highest form of activity. Thus Archimedes, despite his accomplishments, repudiated engineering in the third century B.C. because it was only concerned with utility and profit.[56]

The fact that the Greeks had slaves to whom they could assign the most onerous manual activities was another factor which contributed to the philosophers' lack of esteem for them. Lewis Mumford has maintained, however, that it was this very contempt for the servile nature of the manual arts that provided an incentive for the Greeks to invent, sometimes with the aid of geometry, mechanical, laborsaving devices.[57] In response to his assertion, one could argue that the development of mechanical devices to replace human workers or render their physical labor easier or more productive was a phenomenon driven more by utilitarian and economic considerations than by prejudice against manual occupations.

By Mumford's assessment, in any case, "most of the components of later complex machines were either invented by the Greeks, between the seventh and the first century B.C., or were manufactured with the aid of machines and mechanical parts the Greeks first invented."[58] Key Greek inventions of this Iron Age period cited by Mumford included the screw and its application to the task of raising water, critical for irrigation and instrumental in the opening up of new areas in North Africa and the Middle East to agriculture; suction and force pumps, also useful for water-dependent tasks; the lathe, essential for making accurately turned and bored cylinders and wheels, as well as for producing lifting devices, pulleys, winches, and derricks; metal-stamping processes for coins, which revolutionized economic transactions; and the watermill, which enlisted nonhuman energy for grinding operations.[59]

Not only did the fabrication of these mechanical devices involve a form of *techne*, but so did their use. Plato, as we have seen, emphasized social responsibility in the practice of *techne*, and Aristotle developed a more elaborated theory, integrating it into his doctrine of the four causes. They did not, however, explicitly address in their theory of *techne* a possibility that today is recognized as important—that is, that mechanical devices themselves, by their very design, may impose a certain use on their user regardless of his or her moral and social concerns. Perhaps it was such a realization, in addition to the fact that operating machines required less intellect than professional or philosophical activities, that provided the basis for their belief that manual activities were inferior and could be degrading.

Notes to Chapter 1

1. Camille Paglia, *Sexual Personae* (New York: Vintage Books, 1991), 5.

2. Ibid., 5.

3. Ibid., 5–6.

4. Bertrand Russell, *A History of Western Philosophy* (New York: Simon and Schuster, 1972), 27.

5. Werner Heisenberg, *Physics and Philosophy: The Revolution in Modern Science* (New York: Harper Torchbacks, 1962), 63.

6. Michael Grant, *From Alexander to Cleopatra: The Hellenistic World* (New York: The Macmillan Publishing Co., 1990), 238, 240.

7. Arthur O. Lovejoy, *The Great Chain of Being* (Cambridge, Mass.: Harvard University Press, 1978), 46.

8. Ibid., 58.

9. Russell, 169.

10. Plato, *The Republic of Plato*, trans. Francis Cornford (London: Oxford University Press, 1941), 349–50, 353–54; Russell, 206–7.

11. Paglia, 8.

12. Grant, 235.

13. Russell, 27, 44, 55–56.

14. Plato, *Republic,* 248.

15. Ibid., 250.

16. Ibid., 139–43.

17. Ibid., 119–20.

18. Ibid., 32–33.

19. Ibid., 287–88.

20. *The Encyclopedia Britannica* (1911), vol. 19, 369; *The Encyclopedia of Philosophy* (1969), vol. 5, 467; *Grande Larousse Encyclopédie* (1963), vol. 7, 711.

21. Aristotle, *The Pocket Aristotle*, trans. W. D. Ross, ed. Justin D. Kaplan (New York: Washington Square Press, 1961), 190.

22. Ibid., 227–29, 231–33.

23. Ibid., 263–65.

24. Ibid., 167–68.

25. Hannah Arendt, *The Life of the Mind: One Volume Edition* (New York: Harcourt, Brace, and World, Inc., 1968), Part 2, 16–17.

26. Thucydides, *Heritage of Western Civilization* (Seventh Edition), vol. 1, ed. by John Beatty and Oliver Johnson (Englewood Cliffs, N.J.: Prentice Hall), 94.

27. Jaeger, Werner; *Paideia: The Ideals of Greek Culture*, vol. 1 (New York: Oxford University Press, 1967), 27, 145.

28. Ibid., 145.

29. Paglia, 85, 87.

30. Ibid.

31. Platon, *Oeuvres Complètes*, vol. 1 (Paris: Gallimard, 1950), 90.

32. P. Commelin, *Mythologie Grecque et Romaine* (Paris: Classiques Gallimard, 1960), 308–9.

33. Plato, *Republic,* 22–23.

34. Elspeth Whitney, *The Mechanical Arts in the Context of Twelfth- and Thirteenth-Century Thought* (Ann Arbor: University Microfilms International, 1985), 45.

35. Clarence J. Glacken, *Traces on the Rhodian Shore* (Berkeley: University of California Press, 1967), 54.

36. Whitney, 61.

37. Aristotle, 26.

38. Ibid., 227.

39. Ibid., 26–28, 31–32.

40. Ibid., 30, 32.

41. Ibid., 31–2.

42. Carolyn Merchant, *The Death of Nature* (San Francisco: HarperSanFrancisco, 1989), 31.

43. Aristotle, 110, 112.

44. Whitney, 64.

45. Plato, *The Dialogues of Plato,* vol. 2, trans. B. Jowett (New York: Random House, 1892), 66.

46. Helen S. Lang, *Aristotle's Physics and the Medieval Varieties* (Albany, N.Y.: State University of New York Press, 1992), 124; Russell, 205.

47. Russell, 205–6.

48. Lang, 57, 81.

49. Ibid., 79.

50. Herbert Butterfield, *The Origins of Modern Science, Revised Edition* (New York: The Free Press, 1957), 15–16.

51. Russell, 206–7

52. Ibid., 162–63.

53. Whitney 50–51.

54. Plato, *The Collected Dialogues of Plato, including the Letters*, ed. Edith Hamilton and Huntington Cairns (Princeton: Princeton University Press, 1953) 215, 219–20, 227.

55. Plato, *Republic,* 237.

56. Whitney, 50.

57. Lewis Mumford, *The Myth of the Machine: Technics and Human Development*, vol. 1 (New York: Harcourt Brace Jovanovich, Inc., 1966), 244–45.

58. Ibid., 245.

59. Ibid., 245–46.

2
The Patristic and Medieval Periods of Christian Culture

Christian Ideological Structures Related to Technology and the Technological Ethos of the West

THE ANTHROPOCENTRIC NATURE ETHIC

In contrast to the Apollonian Greeks, whose philosophy did not encourage humans to challenge the overwhelming preponderance of nature in the human-nature relation, the ideological structures of the Christian religion encouraged humans to adopt an activist role vis-à-vis nature by justifying mastery and dominance over other living species, engage in efforts to "improve" on the "given" of the created world of nature, and undertake a collective project to shift the balance in their favor. Thus, whereas Apollonian Greek culture provided the philosophical framework for a static human-nature relation and a theory of a *techne* of limits, Christianity passed on to the West a number of ideas that laid the foundation for a more dynamic relationship and an eventual culture of technology without limits.

According to the British historian Lynn White, the anthropocentric nature morality of Genesis was instrumental in legitimizing the aggressive and domineering attitude toward nature that characterizes the West and underlies its technological orientation.[1] That is, according to the account of the Creation in Genesis, after God created the universe, humans in his own image, and all living things in six days, he gave humans mastery over every living thing on the earth, including all plants and animals, declaring: "[L]et them have dominion over the fish of the sea, and over the fowl of the air, and over the cattle, and over all the earth . . ."[2]

Similarly, the account in Genesis of how God, after putting Adam in the Garden of Eden, requested that he name all the animals[3] can be interpreted as a biblical indication that humans were to have dominion over other living creatures. This idea was reiterated elsewhere in the Old Testament. For example, Psalm 8 said of God: "Thou madest him [the human] to have dominion over the works of thy hands; thou hast put all things under his feet."[4] And Psalm 115 declared that "[T]he earth He [God] hath given to the children of men."[5]

White believed that what he called a species-arrogant and domineering nature ethic was an important cultural determinant of the eventual preeminence of the West in science and technology, he even suggested that the roots of the contemporary ecological crisis could be found in it.[6] St. Augustine, the leading Christian theologian of the early Middle Ages, arguably provided support for this thesis when he wrote, in the fifth century, that humans would prefer non-sentient beings in their world to sentient ones:

> So strong is that preference that had we the power, we might abolish the former from nature entirely . . . sacrificing them to our own convenience. Who, for example, would not rather have bread in his house than mice, gold than fleas?"[7]

Cultures that did not have an aggressive, anthropocentric nature ethic failed to attain the advanced level of technological development that prevailed in the Christian West after the Scientific and Industrial revolutions had been completed. One important example was Chinese civilization, in which a more passive nature morality influenced by Buddhism prevailed. The Buddhist emphasis was not on humans' domination of nature, but rather on their seeking to harmonize and live in balance with it. Each individual, according to the historian Lewis Feuer, was supposed to nullify and absorb himself in nature, to merge into its "great oneness."[8]

It should be mentioned, however, that the Chinese did invent the very technologies that Francis Bacon cited in his *Novum Organum* (1620) as transforming human existence at the time—the compass, gunpowder, and mechanized printing. The fact that they never exploited these inventions as fully as the Europeans did and failed to maintain the technological superiority they enjoyed in the Middle Ages, however, suggests that the Buddhist nature ethic exerted a moderating influence on their technological pursuits and aspirations.

Native Americans, who had a careful, respectful nature ethic, provide an example of a culture without a vocation for technological development. Significantly, they adhered to a biocentric ethic which put all living things on an equal level and discouraged human species arrogance and the domination of nature. Speaking of his own people, Lame Deer, a Lakota Sioux Indian medicine man of the twentieth century, said: "When we killed a buffalo . . . [w]e apologized to his spirit . . . honoring with a prayer the bones of those who gave their flesh to keep us alive . . . praying for the life of our brothers, the buffalo nation . . . To us, all life is sacred."[9]

In the thirteenth century, St. Francis of Assisi, who founded the Franciscan order, made an egregious but ultimately unsuccessful deviation from the official Christian nature ethic of mastery and domination in the Middle Ages. When he advocated human communication with nature and attempted to humanize non-human forms of life, even reminding animals of their obligations to God, he was expressing views with biocentric overtones. He addressed a sermon to the birds and praised the Creator in them, as well as expressing his love for all plants, animals, and for "Brother Sun" and "Sister Water."[10]

It was not surprising, however, that St. Francis's extension of Christian love to other natural species was unpalatable to most of the clergy. St. Bonaventure, for example, who joined the Franciscan order in 1243 and eventually became its general, remained true to the orthodox, anthropocentric view. He asserted that the whole corporeal world, including animals, was created to serve humanity.[11] Furthermore, he emphasized that the creatures of the world were not important in themselves, because their greatest utility was to glorify God and contribute to His felicity.[12]

THE FINISHING/FURNISHING-OF-CREATION THEME

Another strain favorable to technology in the Christian view of nature was the idea developed by Patristic theologians and reiterated centuries later by St. Thomas Aquinas that the Creation was unfinished and that humans, who were made in God's likeness and possessed the faculty of reason, had the exalted role of completing and even improving it. Biblical support for this belief could be found, according to Glacken, in Genesis 1:28, where God instructed humans to "[b]e fruitful, and multiply, and replenish the earth and subdue it"; and in Genesis 2:15, which explains that God put man in the Garden of Eden and then told him to "dress it and keep it."[13]

St. Basil, the chief organizer of monks of the Eastern Church, was in the fourth century one of the first Patristic thinkers to emphasize a human role of completing and furnishing God's creation. He pointed out that the earth's waters were not yet properly confined and argued that technological artifices like dams, dikes, and canals were necessary.[14] St. Ambrose, bishop of Milan, likewise emphasized the improving and furnishing theme. He saw humans working to improve the earth in partnership with God, and he said that the earth was more beautiful when furnished by human improvements.[15]

Illustrative of the finishing/furnishing-of-creation theme also was Cosmas Indicopleustes' description in *Christian Topography* (sixth century) of nature as a house built by God, who then assigned humans the task of completing and adorning it. This idea suggested, as Clarence Glacken put it, that there was no incompatibility between divine purposes and the workings of humans on nature.[16] Although Cosmas and other theologians who articulated it had in mind labor-intensive activities like quarrying, forest-clearing, land-draining, and the planting of vines,[17] the idea that human modification of nature was a God-assigned task provided implicit encouragement for technological development.

Whereas the species-arrogant Judeo-Christian nature ethic entered into Western culture as a permanent ideological structure, however, the finishing/furnishing theme was influential primarily during the Patristic and Medieval periods. The idea of a unique human role to alter and improve the world of nature survives today, however, in a related, secularized version, often used to justify technology without limits. The argument is that whatever humans do to modify and manipulate the phenomenal world of nature with their technology is "natural"—and therefore acceptable—because they are naturally endowed with the faculty of reason, which makes it possible for them to do so. In other words, the possession of reason has replaced the idea in Genesis 1:26 that humans were created in God's image as a justification for assigning them a transforming role vis-à-vis nature.

Another version of this argument is now employed by some apologists for biotechnology to justify the transfer of genetic characteristics from one unrelated species to another, which critics frequently characterize as a radical violation of nature. According to the biotechnologists, however, because humans themselves are a product of nature's evolutionary process, they do not violate nature when they "participate" in this process by manipulating the genes of living organisms to create new life forms.[18]

THE COUNTERVAILING INFLUENCE OF THE *CONTEMPTUS MUNDI* AND END-OF-THE-WORLD THEMES

Competing in the Middle Ages with the activist stance toward nature, which the Christian anthropocentric nature ethic and the finishing/furnishing theme encouraged, was a pronounced otherworldly strain in Christian thought. With the barbarian invasions of Germanic tribes and the shock of the end of Roman civilization in the fifth century, this point of view, which encouraged a passive human stance vis-à-vis nature and provided little justification for technological development, gained in influence. According to Thomas Goldstein, it defined the world of phenomena as a poor reflection of the life of a transcendental world, "asked for the complete denial of the world of the senses," and stamped the medieval world with inwardness and an outlook of stoic acceptance.[19]

Biblical support for such a negative and pessimistic view of the natural world came primarily from Genesis 3:17, which contains God's stern declaration to Adam after the Fall: "[C]ursed is the ground for thy sake." This passage, according to Glacken, was the scriptural basis for the *contemptus mundi* attitude, which he identified as one of the primary ideological currents of medieval Christianity.[20]

Goldstein's characterization of the medieval Christian world as won over by the *contemptus mundi* outlook, however, failed to take into account a countervailing and more positive interpretation of nature that encouraged physico-theology and nature study in the Middle Ages and discouraged excessive otherworldliness and rejection of the world.[21] Biblical support for this viewpoint came from Genesis 1:25, where God declared after his Creation that "it was Good," and from Psalm 104, which praised the beauty and bounties of nature and exulted: "Lord, how manifold are thy works! in wisdom hast thou made them all; the earth is full of thy riches."[22] In contrast to *contemptus mundi*, this view of nature, like the finishing/furnishing theme and the Christian nature ethic, was compatible with technological development and with active efforts to procure the material goods of existence.

Significantly, the disparaging view of nature expressed by St. Augustine, the greatest and most influential in the early Middle Ages of the Christian thinkers, was not without nuance. "For the sky and the earth and the waters and the things that were in them," he said, ". . . are not evil . . . it is evil men who make this evil world."[23] He was careful to add, however, that God was glorified not by the utility and convenience of his creations for human beings, but by their excellence and purpose for their own sake. Furthermore, Augustine made it clear that humans should never forget that the Creator was superior to his creations and that the natural order was lower than the divine order.[24]

The pessimistic idea that the world would eventually be destroyed by divine intervention was another Christian ideological factor which, like the *contemptus mundi* view but in contrast to the finishing/furnishing theme and the anthropocentric biblical nature ethic, provided no incentive on its face for humans to shape and improve nature. Early church fathers of the third and fourth centuries, such as Eusebios, St. Ambrose, and Lactanius, echoed the New Testament idea that the earth would soon be destroyed.[25] And Augustine, who shared the belief, pointedly quoted the doomsday prophecy of Revelation, 20:16: "[. . .] for the first heaven and the first earth have passed away; and there is no more sea."[26]

Despite this end-of-the world theme and the influence of the *contemptus mundi* outlook in Christian thought, Glacken concluded that the dominant idea of the Middle Ages was a belief that humans were endowed with the facility of work and were intended to assist God in the improvement of nature, even if they were only temporary residents of the earth.[27] Indeed, the fact that new agricultural technologies were developed from the sixth to the ninth centuries in northern Europe and the phenomenon in the later Middle Ages of a little industrial revolution based on nonhuman energy sources like wind and water provide concrete evidence that the otherworldly current in Christianity, hostile as it may have been to science, was not enough to stifle technological progress during the Medieval Period.[28]

In thirteenth-century Europe, following the rediscovery of Aristotle's writings and their translation into Latin, Thomas Aquinas worked out an official synthesis of Christian theology and Aristotle's natural philosophy. Not surprisingly for a thinker influenced by the Apollonian Greeks, Aquinas rejected the claim of Origen, the most influential Church theologian to precede Augustine, that nature was a reflection of sin. He maintained, rather, that as a creation of God it reflected His glory and goodness.[29] He also made it a point to condemn explicitly the most egregious example of the *contemptus mundi* view of nature, that of the Albigensian heretics of France.[30] Members of the Manichaean Cathari sect, which was eliminated by the Inquisition in the next century, the Albigensians saw the world as a duality of spirit and matter in which all forms of matter were evil.[31]

THE IDEA OF HISTORY AS A LINEAR PROCESS LEADING TO REALIZATION OF HUMAN DESTINY

In contrast to the cyclical conception of history of the pagan Apollonian thinkers, the Christian and Hebraic view of human time in the world was linear and teleological. This was undoubtedly one of the most important contributions of Judaism and Christianity to the Western intellectual tradition. With the possible exception of the anthropocentric Judeo-Christian nature ethic and the idea that humans were made in God's image, it was also the most important Christian ideological contribution to the eventual development of a philosophical rationale for technology without limits. This influence is obvious when we take into account Francis Bacon's linkage of scientific power knowledge to the onset of a millennium in the seventeenth century and when we consider the Enlightenment *progressistes*' secular theory of constant human improvement over time, spearheaded by scientific and technological innovation and guaranteed to have, without exception, a positive effect by the theodicy of *laissez-innover*.

In Genesis, the description of God's creation of the world for humans, as well as the story of Adam and Eve and the Fall, represent a rectilinear pattern of unique occurrences. The same can be said for the life of Christ, which included the singular events of the incarnation, the crucifixion, and the resurrection, all related in the New Testament. As Paul said in his Epistle to the Romans, Christ died only once for our sins.[32]

Augustine was the most influential Christian father who removed history from the cycles of nature, where the Greek and Roman thinkers had situated it. In his *City of God*, he gave it a linear direction and made it a product of human action. The biggest problem, however, was how to reconcile his emphasis on history with the goals and ideals of the otherworldly Christian sky cult. He did this, essentially, by presenting the movement of history on earth as a fulfillment of God's will, leading to the ultimate realization of human destiny in the heavens.

In *City of God*, written after northern barbarians had sacked Rome in 410 A.D., Augustine described history as an immanent and necessary developmental process, proceeding by stages toward an ultimate goal. He saw this process as characterized by a constant struggle between two conflicting segments of humanity, those who belonged to the corrupt city of man, and those who belonged to the spiritual city of God. The final goal, or purpose, of history was the triumph on earth of the city of God. At that point, Augustine said, God would destroy the world, and there would be a second resurrection of the bodies of the just, who would be gathered in the kingdom of God in the sky.

To combine in Christian theology the contradictory ideas of human possessors of free will making history and an all-powerful God who controlled all the events of the world, Augustine had recourse to the concept of foreknowledge. That is, he said that God knew ahead of time and approved of everything humans did. Thus, by means of this ambiguous formula, Augustine contrived to present history as the product of human free will while describing it as an expression of divine will.

The prophecy in the New Testament Book of Revelation of St. John of a Second Coming of Christ followed by a millennium (1,000 years of betterment) and the end of the world, was also a linear formulation of the movement of history. It, too, included the idea that history had a definite purpose, or goal. Although Augustine had declared that Revelation had only an allegorical meaning and the Church Council of Ephesus officially condemned millenarian views as heresy in 431, the twelfth-century writings of a Calabrian abbot, Joachim of Fiore, reactivated belief in the millennium among some members of the Christian community, including radical monks belonging to the Franciscan order in the thirteenth century.[33]

The idea that the redemption of humanity would take place in the context of the historical process, in any case, was a fundamental Christian belief. It gave the passage of time itself a value which did not exist in pagan thought. When Western thinkers, both Christians and their secular Enlightenment successors, linked technological advancement to the passage of time within a broad historical perspective of inevitable betterment, the idea that technological innovations had to be evaluated in terms of their more immediate and particular effects was diluted. By the twentieth century, some secular thinkers were presenting technological newness and change as ends in themselves, regardless of concrete effects.

Claude Lévi-Strauss, the French anthropologist, clearly understood the necessary connection between cultures that viewed themselves as "making history" and the dynamic of technological development. He referred to those societies that "internalize the historical process as the motive force of their development" as "hot" societies. Those that attempt to "annul practically automatically the effect of historical factors on their equilibrium and continuity" he characterized as "cold." In the latter type of society, many of them tribal, Lévi-Strauss said there was—and is—a determination to retain the technologies of the past for the legitimizing reason that they were handed down by its ancestors.[34] In "hot" societies like those which developed within Christian civilization in Europe, however, technological innovation was—and is—considered to be a crucial factor in the historical process. In the modern world, it eventually became a primary means of human self-affirmation and a source of pride, independent of any concern for actual historical outcomes.

A cyclic sense of events emphasizes the ideas of renewal, connection, and continuation, whereas a linear conception like the Christian one carries the implication that everything is nonrepetitive and has a one-time purpose or outcome, even if it is part of a long-term process. Paglia linked this linear mind-set to the West and to a singularly masculine perspective, ultimately reflective of the nature of male sexuality.[35] It typifies, in any case, the narrow and incomplete approach of contemporary technologists to reality. Not only are the latter no longer interested in historical outcomes, but they believe that technological actions are isolated actions whose validity depends only on immediate consequences.

The a priori guarantee provided by Christian thinkers like Augustine and the millenarians that the forward motion of time would be beneficial carried a danger that the past would be seen primarily as something to be transcended. Such a devaluation of the past could result not only in the forgetting of its lessons, but also in a loss of the goals and values that have guided humanity through troubles and triumphs. Lacking this wisdom and experience, how can humans responsibly shape their future, particularly today, in a world where the Christian God is silent, philosophy has abdicated its role in relation to values, and fallible humans are armed with technologies of incredible power and sophistication?

Evolution of the Christian Theory of *Techne*

PATRISTIC VIEWS: THE IMPORTANCE OF POSITIVE PAGAN THEMES OF *TECHNE*

A primary classical influence on patristic thinkers who commented on *techne* was the Prometheus myth. It is worth noting that the Christian idea of original sin as disobedience to God, related in the account of the Fall in Genesis, has some similarity to the portrayal in the Prometheus myth of the theft and human acquisition of *techne* as a rebellion against the gods. Prometheus was condemned in the Greek myth to suffer through the punishment of the liver-picking vulture, and humans were punished by means of a release into the world of the evils in Pandora's box. In the Christian story of the Fall, all humans were condemned to suffer for the sin of Adam and Eve, most drastically through the divinely mandated punishment of death and the rupture of the harmonious relation with nature Adam and Eve had enjoyed in the Garden of Eden.

An important difference between the story of Prometheus and the story of the Fall was that in the biblical treatment, technology was not stolen from the gods but rather was given to humans by God. As stated in Isaiah, God "instructed" and "taught" humans how to plow the fields, sow seeds, and harvest crops like barley, rye, corn, and wheat.[36]

Cicero, a Roman Stoic living in the century before the birth of Christ, wrote *De natura deorum*, another important classical influence on Christian theologians of the Patristic period. It prefigured, in particular, those patristic thinkers who associated *techne* with the species uniqueness of humans and their ability to create a better world for themselves. In contrast to some of the patristic commentators on *techne* who followed him, however, Cicero gave it an unmitigated endorsement.

Humans were the unique possessors of *techne*, he said, because of their natural endowment of hands, reason, and an upright stance. They alone, he added, were able to control the winds and the seas with the techniques of navigation. They, too, were the ones who sowed corn, irrigated the soil, and confined rivers or diverted their courses. Because of *techne*, he concluded, they were able to create for themselves a "second world" within the world of nature.[37]

Origen, who belonged to the category of patristic thinkers whose treatment of *techne* reflected classical humanism, wrote in the third century that God had made humans more needy in their natural endowments than animals but had endowed them with reason. These factors led to the discovery of the mechanical arts, he added, by means of which human nourishment and self-protection were rendered possible.[38] This idea that physical weakness and limitation were the basis of the human need to develop technology was similar to the justification in Greek mythology for Prometheus's theft of *techne* from the gods.

Theodoret, Bishop of Cyrus (Syria) in the fifth century, echoed themes that had been voiced by Cicero. God, he said, had given humans the intelligence to invent tools and guide their hands and arms in such activities as mining and agriculture. He reasoned, therefore, that thanks to this kind of human artifice, they not only lived, but they lived well.[39] Thus, like Cicero's assessment in *De natura deorum*, he associated technology with human superiority and species uniqueness.

THE SHIFT TO AN AMBIGUOUS VIEW OF *TECHNE*, TYPIFIED BY AUGUSTINE

Other influential early Christian fathers who commented specifically on *techne* combined classical themes with a cautionary view based on the doctrine of the Fall. Nemesis of Emesa (Syria), for example, said in the fourth century that the mechanical arts originated in human indigence and nakedness. He warned, however, that their cultivation could undermine salvation. Eating from the tree of knowledge, he said, had been forbidden in the Garden of Eden because humans would have become so conscious of their physical weakness that they would have devoted all their concern to care for the body, thus neglecting care of the soul.[40] This idea of a limited role for technology not only reflected the sinfulness of humanity, but it was also consistent with the prejudice of Greek thinkers like Plato and Aristotle against the manual arts.

Gregory of Nyssa, of the same period, said that human mastery over iron gave humans the protection that horns and claws provided for animals.[41] Similarly, it was because their bodies were deficient, he said, that humans had acquired dominion over animals, which in turn supplied them with labor, wool, leather for armor and shoes, and feathers for arrows. Like Nemesis, however, Gregory was unwilling to give technology an unqualified endorsement and restricted its acceptable categories to weaponry and animal husbandry.[42]

Given his pessimistic view of the human condition and his very limited acceptance of anything worldly, it was not surprising that Augustine, the greatest of the church fathers, expressed only a half-hearted acceptance of technology in his *City of God*, written in the fifth century. Although he believed humans were depraved due to original sin, he stated that the Creator had conveyed on them countless blessings with which they could alleviate their misery in this world. The body, he said, with its erect stature and heavenward look, was designed in the service of a rational soul. The mind made it possible for humans to understand what was good and true and to acquire virtues, which were a guide to living. The mechanical arts, too, served a consoling purpose and were, therefore, of limited use. They were part of human knowledge, but they belonged to a lower, utilitarian order that did not in itself guarantee humans' ability to employ them for good ends.[43]

Some technologies, Augustine said, in making a distinction as to their merits, were necessary, whereas others were merely for pleasure. Furthermore, although innumerable arts and skills had been discovered and perfected by the natural (i.e., God-given) genius of humans, there were some whose purposes might seem "superfluous, perilous, and pernicious." Not only was technology of no aid to salvation, therefore, but in Augustine's view, it could even be immoral or dangerous when it served primarily to make the burden of human existence less onerous.[44]

This qualified acceptance of *techne*, according to Elspeth Whitney, was deliberately written by Augustine as a response to Cicero's wholehearted endorsement of it.[45] In contrast to the Roman philosopher, Augustine did not consider all technologies to be an expression of human power and dignity. He specifically mentioned agriculture, navigation, hunting, cooking, and ceramics as technologies whose purposes were generally recognized as beneficial; but he also mentioned poisons, weapons, and equipment used in wars as technologies that could be included in a negative and harmful category.[46] Thus, the technology of Augustine definitely did not correspond to the Apollonian idea of neutral means. Rather, it was characterized by moral ambiguity and therefore had limited legitimacy.

It is also important to note that, in contrast to later medieval Christian thinkers, Augustine did not write about technological progress as a way for humans to recover a form of power and knowledge they had enjoyed in Eden before the Fall. Significantly, Jacques Ellul, a twentieth-century French Augustinian who wrote critically about modern technology, has made a distinct effort to make it clear that technology had no place in prelapsarian Eden and did not come into the picture until the events of the Fall.

Ellul pointed out that as a result of his transgression, God said to Adam, as related in Genesis, "Cursed is the ground for thy sake" and "[I]n the sweat of thy face shalt thou eat bread." And to the serpent, He said, "I will put enmity between thee and the woman . . ."[47] Ellul explained that these divine pronouncements mean that the harmony between humans and nature that existed in Eden was broken, that nature was degraded, and that humans henceforth had to work to obtain from it what was needed for survival. At this point, Ellul added, technology, like work, became necessary. Henceforth, he said, technology would be essential for human self-preservation.[48]

The first act after the Fall, according to Ellul, was a technological act performed by God—the fashioning of clothes for Adam and Eve so they could cover their naked bodies and repress sexual lust. The Bible then spoke clearly about technology, Ellul added, emphasizing the postlapsarian origin of technology by mentioning that Adam's son Cain was a "tiller of the ground" and that his brother Abel was a "keeper of sheep."[49]

LINKAGE IN THE LATER MIDDLE AGES OF *TECHNE* WITH POSTLAPSARIAN RECOVERY

By this end of the fifth century, work was taking on a new importance, both practical and theological, in the monasteries. Because of the organic link between tools, machinery, and methods with labor, this was the beginning of the emergence in the Christian Middle Ages of a more favorable conception of technology.

In his *Retractationes*, Augustine himself had commented unfavorably on some of the monks at a monastery in Carthage who wanted to live on "oblations of the faithful." He rejected their view, quoting Paul's words in II Thessalonians 3:10: "If anyone will not work, let him not eat."[50] In the Benedictine monasteries, labor soon became a guiding principle as well as a necessary activity. The most frequently quoted rule of the order, established in the sixth century, was Rule 48, according to which "idleness is the enemy of the soul."[51] Thus, the very labor that had been despised as the lot of slaves and associated with the Fall was now encouraged as more than just a penalty for sin. Furthermore, in a monastic milieu of sexual repression, work also served the practical end of providing a positive and acceptable outlet for bottled-up instinctual energy. Not only were prayer and meditation important in medieval monasteries, but the use of the axe, torch, hoe, and plow were as well. Monasteries, which had begun as a retreat from the world, soon were the scene of energetic activity like land clearing and conversion, planting, stockbreeding, etc.[52] Agriculture was the most common form of work, but the monks also engaged in mat- and basket-making and the copying of manuscripts.[53]

The logical corollary to this positive reevaluation of work was the emergence of explicit theological support for technological enterprise and technology development. In the ninth century, John Scotus Erigena, a Carolingian court philosopher, took a decisive step and broke with Augustine's tentative and morally ambiguous view of technology. In his commentary on Capella's fifth-century work, *The Marriage of Philology and Mercury*, Erigena linked the mechanical arts with recovery of original God-like qualities, which, according to Genesis, humans had possessed before the Fall. He argued that knowledge of the useful arts was part of humanity's original endowment and that teaching about them was merely promoting the recall of understanding that was deeply stored in human memory. As individual humans acquired this useful knowledge and put it into practice, he reasoned, they were reforming themselves in the image and likeness of God.[54]

In the tenth and eleventh centuries, the Benedictines, whose rules were imposed on all religious houses in Charlemagne's realm, also had reached a point where they believed in a divine sanction not only for work, but also for technology. According to David Noble, they put the idea of cultivation of the mechanical arts as a spiritual endeavor into practice in their monasteries, "turning their religious devotion to the useful arts into a medieval industrial revolution, pioneering in the avid use of windmills, watermills, and new agricultural methods."[55]

In the twelfth century, as the end of the Middle Ages approached, a number of leading Christian thinkers in the monasteries expressed a similar view. Although Hugh, a theologian at St. Victor in Paris, referred to the mechanical arts as adulterate, he argued that they served the purpose of remedying human physical weakness after the Fall. Just as the theoretical arts (mathematics, physics, and theology) served to correct human ignorance after the Fall, and the practical arts (politics, ethics, and economics) served to correct moral vices, the mechanical arts were, for him, part of a repertory of knowledge and skills that would serve to restore humans to their natural, prelapsarian condition.[56]

The theologians' association of development of the mechanical arts with human restoration of prelapsarian qualities meant, in effect, that *techne* was seen by Christians in a more positive, less equivocal light. Godfrey, another Christian scholar at St. Victor in the twelfth century, argued in his *Microcosmus* that the practical and mechanical arts were necessary for human life. Furthermore, he said, they had a moral dimension. Their origin, he added, was in the Law of Moses; and they helped to direct the irrational part of the soul to proper goals. They would only be adulterate, therefore, if they were misused for pleasure.[57]

AQUINAS'S DISREGARD OF THE RECOVERY THEME AND HIS RETURN TO ARISTOTLE'S *TECHNE* OF LIMITS

Aquinas was an exception to the overwhelmingly positive endorsement of technology made by a number of leading Christian theologians at the end of the first Christian millennium. Indeed, given that his natural philosophy was a return to Aristotle, who saw contemplation as the highest form of human activity, and to the Apollonian strategy of explaining the world of nature in rational and teleological terms, it certainly would have been surprising if Aquinas had assigned any higher meaning to technology. In contrast to a number of leading Christian theologians of the late Middle Ages, at no time did he portray development of the mechanical arts as a means for humans to recover from the Fall. As Whitney explained, he showed relatively little interest in the mechanical arts and, far more than most of his contemporaries, labeled them servile and degrading.[58]

Aquinas remained true, rather, to Aristotle's categorization of the mechanical arts as belonging to the lower category of activities done for some purpose external to themselves, not to the superior category of things done for their own sake, like philosophy. But although he insisted on their servile and inferior character because of their corporeal nature, he conceded that they could have a positive, but limited, role. "It is far more suitable," he said, "for a rational nature, which is capable of endless ideas, to equip itself with an endless catalogue of tools."[59]

Aquinas did reaffirm the patristic idea that humans had the role of completing and improving God's creation but, it is important to note, without explicitly assigning any particular importance to technology as a means for accomplishing the task. He explained this human role in a way that was not inconsistent with Aristotle's theory of final causes. Since God, Aquinas said, did not do everything personally in this world, he sometimes acted through secondary causes. Thus, on the seventh day of Creation, God began a "second perfection . . . directing and moving his creatures to the work proper to them."[60]

Sensitive to the Apollonian concerns for limits, however, he warned that precisely because God had put rationality and some of the divine in humans, they also had to show humility in their divinely sanctioned rule of nature.[61] Thus, Aquinas would have been shocked if he had lived to see that in the contemporary world humans have moved from the task of assisting God to the task of playing God, armed with technologies having life-and-death power, such as recombinant DNA and nuclear weaponry. In our contemporary culture of technology without limits, there is a danger, as Virilio has warned, that humans will *finish off* creation rather than "finish" or "furnish" it in the sense of completion, which Aquinas and his patristic predecessors intended.

The Mutations of Science in the Patristic and Medieval Periods

SUBSERVIENCE OF EARLY CHRISTIAN STUDY OF THE BOOK OF NATURE TO FAITH AND AUTHORITY

Despite a number of ideological structures favorable to technology and technological development in the long run, and despite the positive role assigned to *techne* by Christian thinkers, for hundreds of years Christian theory of nature was largely as useless as Apollonian natural philosophy had been for the acquisition of knowledge suitable for technological application. The effect of its shortcomings was that although unmistakable progress in the arts and crafts—that is, in *techne*—occurred during the period of Christian preeminence in European culture, it took place without any meaningful contribution from natural philosophy.

The fundamental truths of Christianity came, not from reason or experience, as in modern science, but from revelation and authority. Some patristic thinkers, however, did make a limited concession to sensory knowledge and did encourage the study of the "Book of Nature," but only as a way to learn more about God the Creator. They took as a guide what Paul had said in Romans 1:20: "For the invisible things of Him from the creation of the world are clearly seen." This passage provided the scriptural basis for a *theologica naturalis* and physico-theology,[62] according to which the existence of a beautiful and wondrous nature was proof that God existed. This form of nature study, however, engaged nothing more than the ordinary use of the senses as an epistemological tool, unaccompanied by any rational method or theory.

John Chrysostom, one of the greatest Greek fathers of the early Church, wrote in the fourth century that even the poor and illiterate—all humanity without exception, including barbarians—could learn about God by studying the heavens, which were written in a universal language. This idea that knowledge of superior truth (in this case, God's glory) could be found by means of sky contemplation had an obvious affinity with the promotion by Apollonian philosophers like Plato and Aristotle of the study of the heavens to develop a basis for a higher moral life. All that was necessary for a Christian, Chrysostom said, was to see.[63]

Augustine, too, expressed support for study of the Book of Nature as a way to find God: "Some people, in order to discover God, read books. But there is a great book: the very appearance of created things. Look above you! Look below you! Note it; read it . . ."[64] He made it abundantly clear, however, that there was no place for the rational principles and concepts of Apollonian natural philosophy in the Christian understanding of nature. Faith, rather, was the essential factor.

> It is not necessary to probe into the nature of things, as was done by those whom the Greeks call *physici* . . . nor should we be in alarm lest the Christian should be ignorant of the force and number of the elements . . . It is enough to believe that the only cause of all created things . . . is the goodness of the Creator . . ."[65]

Indeed, given Christianity's heavy component of magical elements like the virgin birth of Christ, miracles, angels, wine turning to blood (transubstantiation) in the sacrament of the Eucharist, and Christ's ascent from the dead after the crucifixion, any encouragement of a search for rational explanations of the workings of nature would have carried the risk of opening the door to unwanted questions about its faith-and-revelation-based theology.

The result of these epistemological shortcomings was that the dynamic and rational body of empirical knowledge, which we would today include under the heading of science, and which can be applied technologically, was for the early

Christian centuries nonexistent. Instead, natural philosophy was largely an obscure realm dominated by mysticism, superstition, and ignorance. Medicine, for example, stagnated because of the stifling and negative influence of the Christian doctrine. Augustine attributed all disease to demons; and since illness was connected with sin, the Church counseled prayers and fasting as therapy. Surgery and dissection were forbidden, since to cut into the human body was to tamper with a divine creation.

THE QUALIFIED RETURN TO REASON IN NATURAL PHILOSOPHY AND ITS CODIFICATION BY AQUINAS

In contrast to a dynamic and even exalted role assigned to technology in human affairs by a number of leading Christian thinkers starting in the ninth century, the scientific study of nature was discouraged until the recovery of Greek manuscripts began in the late Middle Ages. The impetus for the release of natural philosophy from the shackles of faith-and-revelation-based medieval theology came, temporally, from the writings of the pagan Greek past. Geographically, it came indirectly from the Middle East, primarily by way of Latin translations made in southern Europe of Arabic translations that had preserved the thought of the original Greek manuscripts.

Christian scholars flocked to Spain in the twelfth century, where they undertook the task of translating Arabic translations of the Greek texts, as well as original Arabic writings on medicine and mathematics, into Latin. In Toledo, Gerard of Cremona made a remarkable contribution to this effort by translating from the Arabic the core of Aristotle's scientific writings, major portions of Euclid's *Elements of Geometry*, and Ptolemy's *Almagest*, a compendium of ancient Greek mathematics and astronomy that dated from the second century A.D.[66]

In Sicily, the Norman conquest of the eleventh century had resulted in a flourishing Christian-Islamic culture. At the court of Frederick II in the twelfth century, Michael Scot translated from the Arabic Averroes's commentaries on Aristotle, which then became part of the curriculum at the newly founded University of Naples. For the next two or three centuries, there was an Averroist movement in Europe which advocated an exclusively rational approach to the investigation of nature.[67]

Typical of the rethinking of nature brought on by the rediscovery of pagan natural philosophy was the comment in the twelfth century of Abelard of Bath to his nephew, who stated that God was the cause of growth of herbs and plants. Abelard replied that there were natural causes at work as well, adding that nature was "not confused and not without a system."[68] In the next century, Albertus Magnus, a mendicant Dominican, provided an example of the influence of Aristotelian teleology on Christian thinkers when he asserted that not only did the world have a purpose, but so did every living organism in it.[69]

Aquinas played a key role in the second half of the thirteenth century, when he completed his influential synthesis of Aristotelian philosophy and Christian theology. He rejected both the position of Augustinians like Bonaventure, a reactionary Franciscan leader who wanted to base everything on faith, and the Averroists, with whom the Dominicans were suspected of sympathizing and who believed that philosophical truth could be based on reason alone.[70]

In Aquinas's view, reason and faith complemented each other, although faith was never contrary to reason and was always superior to it. This meant, in effect, that truth based on reason had to be compatible with Christian dogma. Certain religious truths could be derived by reason, Aquinas said, including the existence of God (i.e., the arguments about a Prime Mover, a First Cause, a Necessary Being, etc.) and the immortality of the soul. They could also be proven by faith, he added, which was essential for the young and ignorant, who had no time for philosophy. Other truths, like the Trinity, the Incarnation, and the Last Judgment, could only be known by faith.[71]

As far as the truths of the workings of the world of natural phenomena were concerned, Aquinas accepted the authority of Aristotle and his reliance in science on a priori first principles, observation, and logic. Christian science of the late Middle Ages thus took on the trappings—and shortcomings—of Apollonian nature philosophy, adjusted at some key epistemological points to give Christian theology the final say in the case of incompatibilities. It had the same disadvantage as Apollonian nature philosophy, in any case, as far as *techne* was concerned. That is, it was unable to yield any theoretical knowledge which would further technological development.

Despite the Scholastics' introduction of an intellectual method at Paris that included the presentation of a thesis and defending it from objections, their disputations took place within the ideological confines of established orthodoxy. There was none of the appeal to the kind of critical reason that had been recognized as a necessary tool for philosophical inquiry by such ancient Greek thinkers as Parmenides, the Sophists, and Socrates. One of the greatest disadvantages of Scholasticism, therefore, was that it left little room for innovative thinking or the questioning of basic premises. The received ideas of Christian dogma and Aristotelian philosophy, thus rarely challenged, effectively stifled scientific inquiry until the seventeenth century, when Bacon called for smashing the four idols of the mind and Descartes performed his famous philosophical exercise of radical doubt applied to everything.

After Aquinas's version of Scholasticism triumphed, Christian authorities elevated a number of his erroneous scientific theories to the status of unassailable truth. These included his principle of the five elements (originally Empedocles' four); his belief (like Plato and Aristotle) that all living things were part of a divinely created, hierarchical chain of being; and his view, incorporated in Ptolemy's celestial system, that the stationary earth was the center of the universe and that planetary orbits and epicycles were circular. In these and other instances, such as Galen's fallacious theory of the circulation of the blood, the support of Church authority for incorrect theories inherited from classical antiquity constituted a significant—although ultimately only temporary—obstacle to the discovery of empirical truth about nature and the development of a science capable of practical, technological application.

Despite these shortcomings, however, Aquinas's Scholastic natural philosophy served the Apollonian purpose of providing a plausible, rational explanation of natural phenomena, making it compatible at the same time with the basic tenets of Christian theology. Viewed from a longer historical and ideological perspective, it served the purpose of facilitating a transition from the obscurantism of medieval science to the superior methodology and positive truths of modern scientific power knowledge.

ALCHEMIST EMPIRICISM: EXAGGERATED GOALS AND MAGICAL ELEMENTS

Perhaps the most egregious challenge, from a perspective of hindsight, to the idea that Christian theology and Aristotelian natural philosophy were the basis for truth about nature came from the alchemists. Their alchemy was a blend of magic and science that had its western roots in Alexandria in the third century A.D. It reentered European Christian culture in the Middle Ages by the same vehicle as Aristotelian philosophy: Arabic texts. A tenth-century monk, Gerbert, who later became Pope Sylvester II, was, according to tradition, the first European to be familiar with Arab alchemist writings.[72] Although Pope John XXII issued a Bull that excommunicated the alchemists in 1317, and the Inquisition burned a number of them at the stake,[73] the influence of alchemy on nature philosophers and scientists persisted. As late as the seventeenth century, for example, both Isaac Newton and Francis Bacon practiced alchemy.

The alchemists, whose activities and doctrines were clandestine, relied on esoteric texts and secretive laboratory experiments to search for a lost process that would enable them to produce what they called the Philosopher's Stone. According to their belief, whoever succeeded in recovering this process would have a reputed basis for transmuting all metals into gold and for procuring an elixir that would heal all illness and promote longevity. Although their goal was nothing more than a fantasy, however, the idea that scientific experiments would yield knowledge that could be applied practically to manipulate or transform nature was to become the very basis of modern science.

Thus, despite their marginality and their exaggerated objectives, the alchemists were important to the development of modern science because they rejected existing authority, searched for knowledge about nature that would serve practical ends, and were the first to intervene directly in nature's workings in an effort to obtain such knowledge. At a time when the Church counseled humans to consult its theologians and authorized books for knowledge about nature, the alchemists took the radical approach of going directly to the things of the phenomenal world to discover empirically its secrets. Once cleansed of its magical elements and exaggerated objectives by Francis Bacon in the seventeenth century, the experimental method of the alchemists provided an invaluable tool for modern scientific inquiry and paved the way for a post-Apollonian human project to master nature by means of technology.

A number of the most notable medieval practitioners of alchemy were members of religious orders. Albertus Magnus, for example, a Scholastic philosopher who studied alchemy for scientific purposes and was the teacher of Aquinas at the University of Paris, belonged to the Dominican order. Roger Bacon, another leading natural philosopher of the thirteenth century, who shared the alchemists' interest in experiments and the transmutation of metals, was a Franciscan. According to Goldstein, a translation of the Arabic *Secret of Secrets* by Philip of Tripoli gave him the inspiration for his surreptitious method of searching for nature's truths.[74] In Bacon's view, scientific experiments were the most effective means to promote advancement in the mechanical arts. They provided "instruments," he said, that humans could use to understand nature more effectively and to recover the means of domination over it.[75]

Bacon not only practiced experiments, however, but he followed the example of Erigena and Hugh of St. Victor in designating the development of the mechanical arts as a way for humans to recover divine qualities they had possessed before the Fall. Furthermore, he typified the more radical elements of the Franciscan order who were attracted to the millenarian perspective of Joachim of Fiore and saw the advancement of the mechanical arts as a way to prepare for the impending arrival of God's kingdom.[76]

All forms of knowledge, including natural philosophy and experimental science, as well as theology and canon law, were for Bacon important aids to faith and part of divine wisdom that had been imparted to humans in the prelapsarian beginning by God. As he saw it, the specific task at hand was to recover and disseminate this knowledge among humans in general. He believed that both the arts and sciences, as well as labor, would help to repair the losses humanity had sustained because of the Fall.[77]

His vision of technological inventions to be derived from his experimental approach was remarkably modern in the sense that he imagined a diversified array of substances and devices, many of which became practical realities centuries later, after the Scientific Revolution had been consolidated and the Industrial Revolution had begun. Thus, when Roger Bacon wrote about the development of flying machines, medicines, burning lenses, incendiary sub-stances, submarines, and self-propelled vehicles,[78] he anticipated the kind of world we are living in today. It is one where technological artifice has gone far beyond the basic requirements of a post-Apollonian project to master nature, as articulated by Descartes and Francis Bacon in the seventeenth century, to become, in the eyes of some, an end in itself. If we applied contemporary terminology to Bacon's vision, we would say that in the context of the thirteenth century, he had a mentality given to science fiction.

Notes to Chapter 2

1. See Lynn White Jr., "The Historical Roots of Our Ecological Crisis," *Science*, 10 March 1967, 1203–7.

2. Genesis 1:26.

3. Genesis 2:19.

4. Psalms 8:6.

5. Psalms 115:16.

6. White, 1203–7.

7. Clarence J. Glacken, *Traces on the Rhodian Shore,* (Berkeley: University of California Press, 1967), 198.

8. Lewis Feuer, "The Scientific Intellectual," in *Psychological and Sociological Origins of Modern Science* (New York: Basic Books, 1969), 253–60.

9. John Free Lame Deer and Richard Erdoes, *Lame Deer, Seeker of Visions* (New York: Simon and Schuster, 1972), 122.

10. Glacken, 214.

11. Elspeth Whitney, *The Mechanical Arts in the Context of Twelfth- and Thirteenth Century Thought* (Ann Arbor: University Microfilms International, 1985), 196.

12. Glacken, 237.

13. Ibid., 152–53.

14. Ibid., 298.

15. Ibid., 299.

16. Ibid., 300–01.

17. Ibid., 173.

18. Jeremy Rifkin, *The Biotech Century* (New York: Penguin Putnam Inc., 1999), 101.

19. Thomas Goldstein, *Dawn of Modern Science* (Boston: Houghton Mifflin, 1980), 55–57.

20. Glacken, 162.

21. Ibid.

22. Ibid., 157.

23. Ibid., 197.

24. Ibid., 196, 198.

25. Robert K. Merton, *Science, Technology, and Society in Sixteenth-Century England* (New York: Harper Torchbacks, 1970), 73–74.

26. Robert Nisbet, *The History of the Idea of Progress* (New York: Basic Books, 1980), 73.

27. Glacken, 162–63.

28. See Lynn White Jr., *Medieval Technology and Social Change* (New York: Oxford University Press, 1962).

29. Glacken, 233.

30. Ibid.

31. Bridgewater and Kurtz, eds., *The Columbia Encyclopedia*, 3[d] Ed. (1968), 41.

32. Romans 6:9–10.

33. David Noble, *The Religion of Technology* (New York: Alfred A. Knopf, 1998), 23–25.

34. Claude Lévi-Strauss, *La Pensée Sauvage* (Paris: Librarie Plon, 1962), 309–13.

35. Paglia, 10.
36. Glacken, 167.
37. Cicero, *De natura deorum; Academia*, trans. H. Rockham (London: G. P. Putnam's Sons, 1933),, 271.
38. Glacken, 297.
39. Ibid., 300.
40. Whitney, 168.
41. Glacken, 298.
42. Whitney, 169.
43. Augustine, *The City of God* (New York: Modern Library, 1950), 850–53.
44. Whitney, 84, 86–87.
45. Ibid., 85.
46. Ibid, 86–7.
47. Genesis 3:15, 3:17, 3:19.
48. Jacques Ellul, "Technique and the Opening Chapters of Genesis" in *Theology and Technology: Essays in Christian Analysis and Exegesis*, ed. Carl Mitcham and Jim Grote (Lanham, Md.: University Press of America, 1984), 132–35.
49. Ibid., 132.
50. Glacken, 304–05.
51. Ibid., 304.
52. Ibid., 302–4.
53. *International Encyclopedia of Social Science*, vol. 9, 417.
54. Noble, 14–17.
55. Ibid., 13–14.
56. Whitney, 152–3, 160–1, 218.
57. Ibid., 188.
58. Ibid., 214.
59. Ibid., 268
60. Glacken, 234.
61. Ibid.
62. Ibid., 162.
63. Ibid., 203.
64. Ibid., 204.
65. Goldstein, 57.
66. Ibid., 109–10.
67. Ibid., 111–13. Glacken, 219.
68. Glacken, 219.
69. Ibid., 227.
70. Bridgewater and Kurtz, 2128.
71. Russell, 454–55.
72. Serge Hutin, *L'Alchimie* (Paris: Presses Universitaires de France, 1971), 42–43.
73. Ibid., 14, 46.
74. Goldstein, 111.
75. Whitney. 273.
76. Noble, 26.
77. Whitney. 216–18.
78. Ibid., 270, 272.

3
The Renaissance
and the Scientific Revolution

The recovery at the end of the Middle Ages of the learning of the great classical thinkers, lost after the fall of the Roman Empire, was the stimulus that generated the dynamic cultural revival of the Renaissance and the scientific revolution that it, in turn, unleashed. This revolution included the adoption of a collective, post-Apollonian human project to master nature in response to a return of the mechanistic model of nature with questionable or nonexistent higher purpose. It meant also the transition from a science of logical explanation to one yielding power knowledge capable of technological application. These changes constituted the "big bang" in intellectual history, which opened the door to modernity and paved the way, ultimately, to today's technoculture, in which technology is no longer subject to higher values and has become its own justification.

The Consequences of the Return of a Mechanistic Nature Paradigm

THE THREAT TO CHRISTIAN TELEOLOGY

The publication in the sixteenth century of Latin editions of Lucretius's *De rerum natura* (first century B.C.) and the appearance of French translations of it in the seventeenth century served to promote a mechanistic nature paradigm of matter in motion that was incompatible with Aristotle's teleological physics of form and matter. This groundbreaking return to the atomist theoretical model was an important aspect of the scientific revolution of the Renaissance. Faithful to Epicurus's atomist view of a universe of chance with an indifferent God, the ideas in the new editions of Lucretius's *De rerum natura* presented a substantial challenge to the prevailing Christian belief in teleology. Marin Mersenne, a leading seventeenth-century atomist philosopher in France, recognized this incompatibility and stated that he did not think atomism could be made acceptable to religion. Guillaume Lamy, lecturing at the Faculty of Medicine at the University of Paris in 1647, generated controversy when he defended the Lucretian view by asserting that God had thrown dice to determine what perfections each creature should have.[1]

Most of the leading thinkers of the time, however, found ways to reconcile, at least temporarily, elements of pagan atomism with Christian theology. For example, Descartes, who was directly influenced by Democritus's atheistic version of the atomist paradigm of nature as tiny particles of matter in motion, included a God in his system as a First Cause that had set everything into motion. Although some of his critics charged that he did not leave a space for God in the phenomenal world because he had said there was no void in it, Descartes explicitly described God as having an invariable will that guaranteed the consistency of (His) laws throughout the world of phenomena.[2]

At mid-century, Pierre Gassendi, a French contemporary of Mersenne, deliberately undertook the task of working out a compromise between the purposeless pagan world of atomism and the teleology of Christianity. He argued that there was a God who had created the universe ex nihilo and ordered the atoms that composed it in a purposeful way. Doubts lingered, however, and after Gassendi's death, some detractors warned that elements in his thought were dangerous to Christianity.[3]

In England, Newton included atomism in his physics and worked out a similar blending of it with Christian belief in God and a purposeful world. He stated that God had initially formed matter into the atoms and had created a universe in which it was evident "order" and "wonderful uniformity," not chance, prevailed.[4] Such a view of the atoms, however, which owed more to theology than to science, was bound to be weakened as rigorous standards of scientific truth were applied more consistently to the task of compiling a valid understanding of the world of nature.

THE NEW UNIVERSE OF "INDETERMINATE" OR INFINITE DIMENSION

The importance of the atomist revival in undermining the Apollonian legacy of purpose and limits that had been reaffirmed by the incorporation of Aristotle's thinking into Scholastic philosophy, was undeniable, but the revolution in astronomy also had a decisive effect. This revolution included a rejection of the Ptolemaic system, which had adhered to Aristotle's theory of circular celestial motion, finite dimension, geocentrism, and the idea of a fifth, incorruptible element in the heavens.

It began with Nicholas of Cusa's assertion in *Docta Ignorantia* (1440) that the earth was not the center of the universe, whose size was "indeterminate."[5] It gathered additional momentum in 1543 when Copernicus, who admitted the possibility of a spatial system extending "indefinitely" beyond the stellar zone,[6] reaffirmed Aristarchus's heliocentric theory (third century B.C.), which removed humans, the most important beings in the created world according to Genesis, from its center. At the end of the sixteenth century, Giordano Bruno carried the idea of indeterminateness further when he stated that the universe was "infinitely populous," implying that it was of infinite dimension.[7]

In the early years of the seventeenth century, Galileo made empirical observations of the heavens with his telescope, which confirmed for him the Copernican theory of heliocentrism and undermined the idea of a unique type of celestial matter. In 1615, he openly supported the controversial heliocentric thesis with his *Letter on Sunspots*.

Newton, who was a bridge in science from the Renaissance to the Enlightenment, incorporated the ideas of a number of key Renaissance astronomers, including Copernicus, Galileo, and Kepler, into his synthesis in the *Principia Mathematica* (1687). He stated that the planetary system was not geocentric but heliocentric; declared that space and time were infinite; and portrayed the universe as a vast, clockwork-type mechanism, built by a God whose active and continuing presence, as he affirmed in response to criticisms that his God was largely "absent," ensured its stability and regularity of functioning.

This new cosmography of Newton and a number of his predecessors had a jarring impact on human consciousness. It meant that the earth was just a speck moving in an infinity of space—a tiny part of a vast, impersonal mechanism, far from its center. These ideas raised new and puzzling questions about the importance and purpose of life, as well as the existence and nature of God. Indeed, even before Newton's cosmological theory gained acceptance in the scientific community, Pascal's famous words in his *Pensées* (1670)—"[T]he eternal silence of infinite space frightens me"—had taken on greater meaning. So had Pascal's statement, "What is man, in the midst of nature?—a nothing in relation to infinity."[8] Just as disturbing, of course, was a realization that infinity could only imply a lack of purpose in nature.[9]

THE PARALLEL IDEA OF A RESTLESS HUMAN NATURE UNWILLING TO ACCEPT LIMITS

Views of external nature held by humans often provide paradigms according to which they pattern their lives or justify their behavior. It was not surprising, therefore, that at a time when humans rethought external nature—that is, outer space—as being infinite, or without limits, the idea that human nature was characterized by a self-assertiveness that resisted limits gained acceptance.

This attitude typified the spirit of the Renaissance and implied a repudiation of the Apollonian ideal of respect for limits, as well as the Christian idea that original sin necessitated the repression of human nature. It served, furthermore, to undermine the idea of absolute truth and objectively valid moral standards, and it was entirely concordant with the search of natural philosophers for knowledge that would give humans greater power over nature.

In the fifteenth century, when the idea of a universe of "indeterminate" dimension first reappeared, Pico della Mirandola declared that humans possessed a nature that was indeterminate and "almost infinite."[10] Of even greater significance, perhaps, was the declaration in the *Discourse on Method* (1637) by Descartes, one of the innovators of the new scientific power knowledge of the seventeenth century, "I am conscious of a will so extended as to be subject to no limits . . ."[11]

In a like vein, the hero of Christopher Marlowe's play *Dr. Faustus* (1593) embodied in literature this new bold attitude of rejecting limits by selling his soul to the devil in exchange for knowledge and magical power. Similarly, Milton's declaration in *Aeropagitica* (1644) that "God gives us minds that can wander beyond all limit and satiety" represented a related concern that humans should be free of any limitations on the publication of their literary and philosophical works.[12]

Pascal was another Renaissance thinker who described human beings as restless by nature and uneasy with limits, although he presented such a tendency as problematical and did not invoke it to legitimize human activity free of moral and social limits. In his *Pensées* (1670), he observed that a person could not remain alone with his thoughts in a room without feeling boredom and existential malaise. He declared, therefore, that men had a constant need for diversions like the conversation of women, gambling, hunting, wars, and various other projects without which they felt an inner nothingness.[13]

According to his understanding, humans were anxious and driven creatures who valued the pursuit of something over its attainment. They needed continual movement and activity, he said, never lived in the present, and were always thinking about the future.[14] These tendencies, it should be noted, are not only concordant with the future-time orientation of the idea of progress, but are also contrary to the Apollonian principle that one's actions should be subject to moral and philosophical limits. It is not surprising that in our contemporary context, some technologists have invoked this very restlessness of human nature as a legitimization for scientific and technological innovation without end.

FORMULATION OF A POST-APOLLONIAN PROJECT TO MASTER NATURE

In the teleological world of Greek Apollonians like Plato and Aristotle and Christian theologians like Aquinas, the primary focus of reason had been to know the idea of the good as a guide for living in the world. When the pagan atomists' version of nature as a mechanism without higher purpose reappeared in the Renaissance and endured despite a number of dubious attempts to reconcile it with Christian teleology, however, thinkers like Machiavelli, Bernardino Telesio, Francis Bacon, and Descartes searched for the epistemological means to make it possible to pursue a new option—that is, to shift from the Apollonian strategy of a philosophical accommodation with nature to the modern strategy of mastering it. At this point they realized, as William Leiss has explained, that there were only two possible choices: either master the forces of nature or be mastered by them.[15]

In other words, the time definitely had arrived for a post-Apollonian science, one with a new method, scope, and goal. It had already been anticipated outside the mainstream of natural philosophy, as we have seen, in the efforts of the medieval and Renaissance alchemists. They anticipated modernity not only by going directly to the physical things and conducting experiments, but also by choosing a goal that involved action—the transformation of nature.

In the seventeenth century, Bacon declared that his new experimental method would yield secrets that would enable the human race to extend its power over the entire universe;[16] and Descartes claimed that with his new subjective philosophy, which included mathematics and experiment when it came down to matters of science, humans would render themselves "masters and possessors of nature."[17] In adopting such a project, these master thinkers were not only taking God's declaration in Genesis that humans were to be the rulers and masters of nature fully to heart, but they were calling upon the human race to adopt the means, in effect, to make it a practical reality. Furthermore, they were expressing on the level of humanity itself the new, modern spirit of self-assertion that had emerged in the Renaissance.

The Elements of a New Science of Nature, Suitable for Technological Applications

CONCEPTUALIZATION: OBJECTIFICATION AND DISENCHANTMENT BY MEANS OF QUANTIFICATION

At the same time the innovators in natural philosophy advocated human mastery of nature, they conceptually neutralized and objectified it. In part, this objectification was the result of a theoretical return to the Democritan idea that the true reality of nature was one of atoms with quantitative properties. When this view gained the upper hand in the seventeenth century, everything was reduced by means of numbers to a kind of radical equivalency that facilitated mastery and domination. When someone protests today, "I don't want to be just a number," he or she is objecting precisely to being stripped of all individuality and specificity for the purpose of administration and control. The quantification of nature has a similar effect of negating the richness and diversity of everything in it in order to submit it to human will. Nature was thus "numbed," numbered, and emptied of its living content and sensual reality by leading natural philosophers of the seventeenth century. Everything, in effect, was made ready for domination.[18]

Both Galileo, in *The Assayer* (1623), and Descartes, in his description of the bowl-of-wax experiment in *Principles of Philosophy* (1644), for example, adopted the Democritan position that sensory perceptions of sounds, tastes, colors, heat, odors, and lightness or darkness were really only the product of the mind and not truly in the object. In effect, these were qualities that scientists at that time did not consider quantifiable. On the other hand, Galileo and Descartes maintained that what stimulated the mind to form these subjective impressions—atom-like particles with mathematically determinable characteristics of size, shape, and motion—really existed *in* the object. In other words, despite the qualitative richness of sensory appearances, such perceptions did not correspond for them to the objective, mathematical reality of nature.

This disenchantment of nature and transition to a quantitative science had been facilitated by the replacement of the cumbersome Roman numerals by Arabic figures, which had been introduced to Europeans, along with the use of equations by Arab mathematicians, at the beginning of the thirteenth century.[19] The translation of Euclid's *Elements* of geometry in the sixteenth century, and the invention of calculus by Newton and Leibnitz at the end of the seventeenth, also contributed to this aspect of the rethinking of nature.

Some European philosophers offered competing—and ultimately unsuccessful—vitalistic theories of nature, according to which it was alive and animated by a spiritual element. In the fifteenth century, the Florentine Neo-platonist della Mirandola stated that wherever there was life there was a soul, and that everything in nature belonged to an interconnected unity of life. In the sixteenth century, Bruno said God was immanent in nature; and in England, Gilbert described the whole universe as animated. Campanella, a seventeenth-century Italian contemporary of Descartes, conceived of the whole world as a living creature.[20] In England, Henry More argued there was an incorporeal spirit of nature that penetrated all matter.[21] And Spinoza, a Jewish philosopher from Holland, expressed the view that God and nature were one.

The claim of these thinkers that the natural world was animated by a spirit, however, was neither ideologically nor practically compatible with the belief that humans should consider nature as an object for their manipulation and control. And given the actual power over nature that the new experimental method promoted by Bacon and employed in physics by the likes of Galileo, Descartes, and Newton promised, the vitalist view of nature was doomed to rejection.

THEORY: THE REDEFINITION OF NATURE AS MATTER IN MOTION, EXPLAINED BY PRINCIPLES OF FORCE AND MATTER

In physics, Telesio of Italy made a crucial departure from Aristotle and his Scholastic successors by returning in the second half of the sixteenth century to the mechanistic theory of the Apollonian nature philosopher Parmenides. In the fifth century B.C., Parmenides had stated that the basic elements of nature were fire, the hot, active factor; and earth, the cold, passive one. Starting with these concepts, Telesio refined them into the principles of external force and matter, thereby breaking with Aristotle's assertion that matter had an inherent active component that moved it toward actualization of its latent form.

In 1565 and 1586, Telesio published in two installments his *De Rerum Natura*, where he rejected Aristotelian authority and argued for a positive approach to knowledge based on empirical observation. His goal was to explain phenomena in terms of their own principles, not external philosophical or theological ones. One should proceed, he said, from the observed to a physical, not a logical, explanation.[22] Bacon, who acknowledged Telesio as a precursor decades later, was definitely thinking along the same lines when he stated that he wanted to govern nature, not in "opinions," but in "action."[23]

Telesio's shift of the nature paradigm from form and matter to force and matter, in any case, was a step toward a science that could yield knowledge useful for acting on the world. His focus on efficient causes and his deliberate disregard of teleological factors helped to redefine physics as the study of matter in motion. In particular, it made it possible to study motion in terms of the action of external forces on material bodies and facilitated a quantitative rather than a qualitative approach. Moreover, Telesio believed, as had the Greek atomists, that there were no determinable higher principles in nature. For him, all phenomenal reality was matter in motion; and existence was characterized by conflict among active forces and a struggle for self-preservation.

One of the most important successors to Telesio was Galileo, who began lecturing at the University of Padua over a half century after Telesio had completed his studies there. Galileo performed experiments with a pendulum and falling bodies at Pisa, and he was one of the Paduan school of natural philosophers who led the way to a modern science that combined hypothesis, experiment, and precise mathematical measurement.[24] In this sense, Galileo was more modern than his British contemporary, Bacon, who advocated adoption in science of the experimental method but did not fully understand the importance of hypothesis nor place sufficient emphasis on quantification. In contrast, Galileo succeeded in expressing a number of physical principles in terms of mathematical laws.

EPISTEMOLOGY: THE SHIFT TO EMPIRICISM

The scientific revolution of the Renaissance not only involved a return to the mechanistic Apollonian nature models and to a quantifying theory of physical reality, but it also was based on the development of an inductive empirical method. These modifications were essential if the new science was to be one that would provide power over nature by means of technological applications. They meant that in contrast to the Christian period, when science, with the exception of alchemy, had no value for technological development, the Renaissance marked the beginning of the linkage of scientific progress and technological advance. This juncture, of course, was a decisive step in the direction of today's technoculture.

Machiavelli's revolutionary use of induction in social science. Although it was fashionable after the seventeenth century to speak of a "lag" in the progress of the humanities and social sciences with respect to advances in the natural sciences, the first steps toward a modern scientific methodology were actually made in political science by Machiavelli, roughly fifty years before Telesio made a similar innovation in physical science. In 1513, Machiavelli published his famous work, *The Prince*, a handbook on the *techne* of power for rulers. His aim was to provide knowledge that would enable a ruler to ensure self-preservation (i.e., acquisition, preservation, and extension of power) in the bloody and treacherous arena of political competition.

Indeed, the political world in which the ruler acted was similar to the world of nature in that higher purpose was absent. Machiavelli was therefore not interested in higher moral principles, only in determining the most effective way to produce a desired outcome. "In the actions of men . . . ," he declared, "one judges by the result."[25] In other words, there were no moral limitations on political actions, which were to be judged solely in terms of their success.

To create a political science upon which efficient technique could be based, he adopted the same premise about human affairs that Descartes and Newton later applied to natural phenomena: That is, he postulated that the course of human history ran in repetitious patterns because human nature was always the same—consistently bad. This approach amounted to a recognition that human events exhibited regularities and were subject to cause-and-effect relations like natural phenomena.

For Machiavelli, half of what determined historical outcomes was the situational factor, *fortuna*. The final result, however, depended on the ability of the ruler to apply successfully his political *techne* to the situation. Furthermore, since events followed consistent patterns, a particular action or tactic that had succeeded or failed in the past would do so again. Machiavelli examined, therefore, the empirical record of history to develop a catalogue of particular situations where famous rulers in the past had either succeeded or failed. He then advised contemporary rulers in similar situations to adopt the tactics that had worked in the past. He thus employed an inductive approach, using specific events to discover general laws of political *techne* a full century before Bacon's *Novum Organum* applied a similar approach to the scientific investigation of nature.

Bacon's plea for induction in physical science. Bacon played the key role in promoting the experimental method in natural philosophy and linking it with the fulfillment of human destiny. He pointedly acknowledged his respect for Telesio as the "first modern," since the Italian had rejected the authority of Aristotle and had advocated going to the "things themselves."[26] Similarly, he praised Machiavelli for writing about what men do rather than what they ought to do, and described him as a precursor, for "drawing his knowledge freshly . . . and out of particulars."[27] "Our only hope," Bacon pointedly wrote, ". . . lies in a true induction."[28]

In *Novum Organum* (1620), he made it clear that his goal was to obtain knowledge that would give humans power over nature. In other words, there had to be a science that would be technologically applicable. "Human knowledge and human power," he declared, "meet in one."[29] To obtain such knowledge, however, Bacon warned that observation alone would not suffice. Nature, he said, had to be "tormented by art."[30] This meant recourse to an experimental method by which humans directly intervened in natural processes by artfully contriving situations designed to reveal knowledge of nature's operations. "Nature to be commanded," he stated, "must be obeyed."[31]

His approach was similar to that of the alchemists, who played a role in developing practical science like pharmacology and believed that power over nature could be obtained by forcing nature itself to reveal its secrets in experiments. Bacon acknowledged their importance as precursors who searched for a single, universal method and rejected authority by going directly to nature and conducting experiments.[32] At the same time, however, he pointed out serious shortcomings in their approach. He derided them, for example, as "a race of chemists" who, "out of a few experiments of the furnace, have built up a fantastic philosophy."[33]

DESCARTES'S EPISTEMOLOGICAL ALTERNATIVE: SUBJECTIVE REASON AS MEANS TO THE MASTERY OF NATURE

Descartes was a Renaissance thinker whose importance in the history of ideas rivaled that of Bacon and Galileo, although his contribution to science was less important than Bacon's, and his influence on philosophy in the long run was undoubtedly greater. We could say that his subjective philosophy, along with Bacon's narrow-focused science, laid the groundwork for a cultural environment in which the power knowledge of the new science would gain the upper hand over higher values and eventually come to supersede them in a civilization of technology without limits.

Whereas Bacon's new method was designed exclusively for scientific inquiry, Descartes's recourse to subjective reason as the basis of knowledge had both scientific and metaphysical applications. He also differed from Bacon because he published his *Discourse on Method* (1637) in French, a vernacular language, in an effort to reach a broader audience than the narrow circle of scholars who knew Latin and could read Bacon's *Novum Organum*.

Like Bacon, Descartes expressed a desire to break out of the epistemological morass where Scholasticism, which he qualified as that "speculative philosophy which is taught in the schools,"[34] had mired European thought. One of the most salient differences between these two innovators, however, was that Bacon's experimental method was inductive and empirical, whereas Descartes's approach, primarily philosophical and mathematical, was rationalistic and deductive. Although he recognized the senses as a "vital" tool upon which humans relied in their daily lives, Descartes did not believe they were adequate for determining the ultimate nature of things.[35] On the other hand, he asserted absolutely that certain metaphysical truths could be derived from the reason of a human subject alone.

In addition to his frustration with the shortcomings of Scholasticism, Descartes's desire to find a new philosophical method that would yield "certainty" was conditioned by a number of other ideological developments in France. These included Montaigne's legacy from the previous century of irony and skepticism, as well as his controversial emphasis on the *moi*—that is, the self—as the only reliable source of knowledge; the unsettling effects on French

Christianity of the sixteenth-century wars between Catholics and Protestants and the seventeenth-century disputes between Jansénistes and the Jesuits, both Catholic; and the publication in France of the writings of the pagan atomist Lucretius, who saw the phenomenal world as devoid of higher purpose.

In his *Discourse*, Descartes explained how having been disappointed by the results of his academic study, and stimulated by foreign travel, he eventually resolved to find a new path to the truth exclusively within his own mind. In other words, he adopted a controversial subjective approach, defiant of church authority, to the acquisition of philosophical truth. As a first and radical step, he deliberately submitted all existing theories, beliefs, and impressions—a broad category that conveniently included the entire body of Scholastic thought—to a process of radical doubt. In the midst of such doubting, however, he found he was able to come up with several "clear and distinct"—and therefore, by his standards, absolutely true—ideas.[36]

The first of these mentally discovered truths was a philosophical verification of the reality of his own existence—and essence—as a *cogito*, a thinking ego. "I think," he declared, as he doubted everything else, "therefore, I am." He followed up this discovery with other truths obtained by means of his mind alone. These included proof of the existence of God; an understanding that the tiniest components of phenomenal reality—"simple natures"—had mathematical characteristics; a knowledge of mathematical laws, which for him were the laws of nature; and an assurance that a perfect God with an invariable will would not alter these laws from one day to the next.[37]

These determinations not only revealed the elements of a post-Scholastic approach to philosophical truth, but they also provided the epistemological basis for a post-Apollonian project to master nature as object. That is, if natural phenomena obeyed consistent, mathematically formulated laws, and if humans were capable of knowing these laws, then they would be able to manipulate and control nature. When Descartes declared that his new, subjective philosophical method would make it possible for humans to "render themselves masters and possessors of nature" and invent "an infinity of artifices,"[38] he obviously had in mind not only the discovery of these laws, but also their technological application.

Michael Zimmerman has offered an interesting psychological explanation for Descartes's project to master nature. According to him, Descartes's solitary "I think" exercise in the *Discourse* led to an attempt to reach God status, with negative consequences for the world of nature. That is, Zimmerman theorized, when a consciousness totally identifies itself with mind—with its *cogito*—as Descartes did, it experiences a radical separation from its surroundings, resulting in fear of what is distanced—that is, nature. This leads, as Zimmerman pointed out, to what Ken Wilbur called the "God project," a vain attempt by the anxiety-ridden ego to render itself omnipotent and immortal by destroying or subjugating what is external.[39] Mediated by technology, this aggression seeks to enhance the subject's security, but in the long run, it threatens to achieve a contrary result by destroying nature, the very material basis of human existence.

Descartes realized, in any case, that in order to determine the practical application of the scientific knowledge obtained by the reason of the thinking subject, a recourse to empirical experimentation was essential. It was necessary, he said, to discover which rationally determined general principles were merely possible and which actually existed in nature.[40] Without such experiments, humans would not have had the practical means to render themselves "masters and possessors of nature." So even though Descartes's rationalist path to the mastery of nature provided an alternative to Bacon's empiricism, it did admit the latter through a back door. And without this concession to experimentation, Descartes would not have been a true instigator in the seventeenth century, like Bacon, of the modern, post-Apollonian project to conquer nature.

THE LONG-TERM PHILOSOPHICAL IMPLICATIONS OF DESCARTES'S SHIFT TO SUBJECTIVE REASON

Although Descartes was a revolutionary thinker in science, the impact of his new subjective method had far greater importance for philosophy. Not only did it amount to a powerful repudiation in France of Scholasticism and its reliance on authority; but in the twentieth century, Heidegger named Descartes as the one who had prepared the way for modern metaphysics, or—to put it in Heidegger's own words—inaugurated the "age of the world picture."[41]

In contrast to the Greek Age, said Heidegger, when humans were "looked upon by that which was," and the Middle Ages, when everything was referred to God and assigned a place in the created order, the age of the world picture was one in which the thinking subject saw all things as external to itself and declared what their true beings were.[42] In other words, Heidegger saw Descartes as the one who opened the door for subjectivity to assume a dominant role in Western philosophy. This was a key step, followed in the next two centuries by such thinkers as Locke, Hume, Kant, and the German idealists. It led eventually to Nietzsche, who announced the death of metaphysics and the emergence of nihilism. Both of these characteristics are reflected in contemporary postmodern culture, which is dominated by technological means rather than oriented by philosophical ends.

The Cartesian transformation of metaphysics, Heidegger added, amounted to the "anthropologizing of being" and a situation in which humans mistakenly believed they could shape their own essences in an unlimited way. Henceforth, he said, they considered themselves able to "give the measure and draw up the guidelines for everything."[43] It was precisely this "anthropologizing of being," as we shall see, that the twentieth-century philosophers of the structuralist and deconstructionist schools, following in Nietzsche's footsteps, associated with humanism and rejected. In doing so, they helped shape a philosophical context in which postmodern nihilism and technology without limits would thrive.

The Characteristics of Scientific Truth After Bacon

THE DIFFERENTIATION OF THE METHODS AND GOALS OF SCIENCE FROM PHILOSOPHY AND RELIGION

Significantly, Bacon's sharpest criticism in *Novum Organum* had been for the church-sanctioned Scholastic philosophy, which he rejected as totally useless for the purpose of obtaining any practical power over nature. He pointed out that at the very time when the investigation of nature was stifled in Europe by reliance on logic and unquestioned first principles, technological inventions like the compass, the printing press, and gunpowder were changing the face of the world. He derisively compared Scholastic philosophers to spiders, spinning webs out of their own interiors; but he made a positive analogy of experimenters to bees, who extracted matter from flowers and refashioned it to serve their own purposes.[44]

Although Bacon was a Christian, he, like Galileo, insisted on the separation of science and religion. He wrote of the "unwholesome mixture of things human and things divine," and warned that "the corruption of [natural] philosophy by superstition and admixture of theology is far more widely spread and does the greatest harm . . ."[45] Furthermore, he considered from his Christian perspective that human invention of a false philosophy like Scholasticism was a kind of second Fall for the human race, similar to the Fall in Genesis, which had cost humans their original domination over nature.[46]

To rectify such errors, Bacon maintained that not only were the epistemological methods and fundamental concerns of religion and science to be kept separate, but in the future, science was also to be purged of any philosophical aspirations. In other words, Bacon wanted natural philosophers to focus on gaining positive knowledge about the functional *how* of the world rather than speculating about the metaphysical and moral *why* of the world.

Like Bacon, a number of other methodological innovators acknowledged that in exchange for power knowledge over nature, humans had to recognize the inability of their methods to know the ultimate truths of the world. Telesio, for example, insisted on observation but said it could provide only knowledge of operations in nature, not the Operator.[47] Thomas Hobbes, who believed that the rational, deductive method of geometry was the "only science humans possessed," readily admitted it was impossible for them to know the final causes of phenomena. He wrote in *Leviathan* (1650) that when humans "calculated the motions of heaven and earth," they did so inside "closets in the dark."[48] Newton expressed a similar point of view when he stated that science provided knowledge about the relation of things, not their essence.[49] Descartes, of course, was the main exception among the innovators in natural philosophy. Like the others, he rejected the Scholastic science of logical explanation, but he presented his new theory of knowledge as valid for both science and philosophy.

The Baconian severing of science from religion and philosophy marked the beginning of the one-sided education of the scientist and his increasing specialization. It meant that science was stripped, as Theodore Roszak put it, of "its ethical vision, its metaphysical resonance, its existential meaning."[50] Furthermore, necessary as this change was for the development of a new science with practical applications, it served to undermine the Apollonian idea that higher, philosophical values should place limits on the use of the productive knowledge of *techne*. And if one day philosophy itself came to rely on only scientific reason for its investigations, the most important philosophical questions would be answered only in relativistic and instrumental terms and metaphysical questions would have to be abandoned entirely.

THE IMPLICATIONS OF THE OBJECTIVE ATTITUDE OF THE SCIENTIST

In the *Novum Organum*, Bacon spoke of the necessity for the scientist to repress any feelings or moral preconceptions that would interfere with his impartial judgment when engaged in the experimental investigation of nature. "The mind," he said, "must itself be from the very outset not left to take its own course, but must be guided at every step, and the business be done as if by machinery."[51] This was why Bacon said in *Novum Organum* that the idols of the mind that he designated as those of the cave and the tribe had to be "smashed." He explained that the former derived from the variability of the "spirit of man" and the fact that each individual's understanding of nature was influenced by his particular education, character, and impressions.[52] And in his commentary on the idols of the tribe, Bacon warned that "affections color and infect the understanding."[53] In other words, Bacon insisted that the scientist maintain, when engaged in his professional activity, an objective attitude, which Nietzsche appropriately characterized as the "incapacity for love."[54] Bacon's scientist had, in effect, to repress all his individual emotions and feelings, including love, pity, and guilt, while engaged in the pursuit of power knowledge that would serve the domination of nature. As Roszak put it, he had to see nature with "the eyes of the dead."[55]

It is important to note that from an epistemological standpoint, however, this neutral, objective attitude was essential. That is, it was intended to ensure that when other scientists followed the same experimental procedure, they would not bias the results by injecting personal judgments or idiosyncratic perceptions into it. Objectivity was a necessary attitude, therefore, to ensure that science would be a collective and international praxis with universally verifiable results. Indeed, if scientific laws were not universally valid, they could not provide the kind of power knowledge that permits humans to predict and consistently control natural phenomena.

The problem is that if this professional requirement to leave out emotions and moral convictions invades and defines a scientist's personal existence, it renders him a dull and drab individual, without a distinctive personality. Similarly, if objectivity prevails in the exercise of power directed at humans or nature, it means not only the absence of bias, but it a failure to have regard for, or even recognize, living beings in their concrete identity and individual differences and needs.

It is significant that Bacon, who instructed scientists to repress their morality and feelings in the pursuit of knowledge, described his experimental method precisely in terms of aggression, saying that the scientist had to "torment" nature in order to obtain its secrets. And although he condemned those who would use this knowledge to enlarge their own power in their own country as vulgar and degenerate, and qualified those who would use it to extend the power of their nation over the rest of the human race as covetous, Bacon described its use to enlarge the power of humans over the entire universe of nature as a sound and noble enterprise.[56]

PIECEMEAL CONSTRUCTION OF SCIENTIFIC TRUTH: BUILDING A "WORLD MODEL" IN BACONIAN TERMS

It is important to note that Bacon said, speaking of the limitations of his new experimental method, "Our purpose is to dissect her [nature] into parts."[57] "I am building in the human understanding," he added, "a true model of the world . . ." by means of a "diligent dissection and anatomy of the world."[58] Such an approach meant that the complexity of nature would be investigated by the Baconian experimenter piecemeal, since his senses could not observe the whole, or even a significant number of different elements of phenomenal reality, at any particular time and in one particular place. The number of variables observed had to be reduced; and practical time and space limitations had to be respected. In other words, empirical truth about the phenomenal world of nature could only come in bits and pieces. Theory, as Descartes explained, could be mathematically derived, but validation would have to be by experiment and would be subject to these same limitations.

In contrast to Bruno, the sixteenth-century Italian who had reasoned that there was an infinity of worlds to be known and that the quest for knowledge of nature would be endless, Bacon optimistically—and mistakenly—believed that within a few decades humans would have assembled all the parts of his world model and that the investigation of nature would be completed. "The particular phenomena of the arts and sciences are in reality but as a handful," he said. Therefore, he added, "the invention of all causes and sciences would be the labor of but a few years."[59]

Not only was Bacon too sanguine in thinking his world model would be completed in a short period of time, but his reductionist experimental methodology did not make it possible to arrive at a correct understanding of nature's operations. In truth, nature consists of complex systems whose components are interrelated and whose most important ecological properties reside in their interconnections, the very factor which his approach excluded.[60] For this reason, technological operations, some applications of Baconian science and others derived independently of it, often ignore these connections and produce both a desired result and unwanted, harmful results—that is, side effects. Such operations, therefore, frequently do not necessarily validate themselves as positive if broader, holistic criteria are applied. This is why modern technology is not only a way to solve problems, but is also the source of an ever-increasing inventory of problems of escalating complexity and seriousness.

THE ABSENCE OF FINALITY IN THE PROCESS OF SCIENTIFIC TRUTH-SEEKING

Bacon admitted that his inductive scientific method was an ongoing, self-correcting process, not a way to discover absolute, eternal truths. In stating that he expected his own data to be corrected by later experiments,[61] he recognized the tentative nature of scientific knowledge. "I am certain of my way," Bacon declared, "but not certain of my position."[62]

His piecemeal, progressive approach necessitated a division of labor that rendered science a collective praxis. In Bacon's view, scientists in different areas were to compare results, and some would seek to verify results by carefully repeating the experiments of others. This meant that scientific truth would henceforth be collectively determined by an international corps of specialized individuals who shared common professional interests.

Since a discovery in one area could lead to breakthroughs in another, or the combination of two discoveries might render possible further advances in the same field, neither the ultimate direction nor the specifics of scientific and related technological progress were foreseeable. Furthermore, this phenomenon of linkage gave an inherent momentum to the process by which scientific knowledge would expand. Bacon referred to this dynamic when he stated: "Axioms duly and orderly formed from particulars easily discover the way to new particulars, and thus render the sciences active."[63] He assumed that this process would cease only when humans had succeeded in building their "world model"—that is, when they had discovered all the laws of nature. Whether such total knowledge would ultimately have been necessary for human well-being and whether it might even have become a threat to the well-being of both humans and nature were questions he failed to consider.

In the twentieth century, Ellul described this same blind process of linkage at work in the dynamic of technological development.[64] It means, in effect, that there are no predetermined limits on scientific research and technological innovation. Not only does each discovery, alone or in combination with others, provide the basis for the next; but a significant number of scientists and engineers would become superfluous if they were not permitted to go on to the next possibility, regardless of the absence of any valid social need for a further breakthrough or the danger of negative consequences.

VALIDATION OF SCIENCE BY UTILITY, NOT HIGHER PHILOSOPHICAL PRINCIPLES

A corollary to the transition to the power knowledge that Bacon's new scientific method provided was that the less humans concerned themselves with questions relating to the origins, morality, and final purposes of the world, the more willing—and able—they were to act upon it and transform it without concern for limits. Scientific truth of the Baconian variety was, in effect, self-validating, primarily by demonstrating its utility in the manipulation and control of nature. Thus, whoever said "science" after Bacon usually implied utility—and technology, meaning its practical application.

This functional criterion implied a rejection of the Aristotelian principle that productive knowledge had to be guided by practical (i.e., political) wisdom and philosophical truth. The whole traditional framework of higher values that was supposed to orient technology and control its use, therefore, was essentially disassociated from it. In this sense, the civilization of modernity, which began with the development of the new, post-Apollonian power knowledge, was increasingly one of practice without a guiding philosophical theory, or, to put it another way, a culture of means without a higher end.

THE NEW INSTITUTIONAL CONTEXT OF SCIENTIFIC INVESTIGATION

Bacon not only forged a link between scientific discovery and technological innovation, but he typified the new institutionalization of the role of scientists when he advocated the creation of state-sponsored research centers. There, in contrast to the solitary and secretive activities of the alchemists, scientists would be paid to serve the interests of the state, and their personal responsibility would be diluted in an organizational framework. Within these institutional settings, however, there were new financial and political constraints that would affect the focus of scientific inquiries and the character of their determinations of truth.

In his *New Atlantis* (1627), Bacon imagined an association of scientists, Solomon's House, dedicated to research and the advancement of scientific knowledge. This idealized conception was the prototype of contemporary government and corporate think tanks, and it prefigured the actual creation of the Academy of Science in France and the Royal Society in England during the second half of the seventeenth century. These learned institutions were followed by the establishment of scientific academies in leading urban locales like Berlin, St. Petersburg, Dublin, Edinburgh, Stockholm, Copenhagen, and Philadelphia in the eighteenth century. With their creation, science became a more public, communal, specialized, and state-sponsored activity.[65]

It was clear, however, that Bacon did not consider the scientists working in his model research center as totally disinterested servants of humanity. He wrote of research fellows in Solomon's House who would spy on scientists elsewhere:

> We have twelve [research fellows] that sail into foreign countries, under the names of other nations (for our own we conceal), who bring us the books and abstracts and patterns of experiments of all other parts. These we call Merchants of Light.[66]

Furthermore, he wanted to place the control of scientific discoveries in the hands of the scientists themselves; and he advised that they keep some inventions secret from the general public, although they might be divulged to the government. He wrote:

> We have consultations which of the inventions and experiences we have discovered shall be published and which not; and take all an oath of secrecy for the concealing of those which we think fit to keep secret, though some of those we do reveal sometimes to the state, and some not.[67]

THE CONTRADICTIONS OF THE PURSUIT OF TRUTH BY CAREER SCIENTISTS: SWIFT'S ACADEMY OF LOGADO AND THE "HEIDELBERG APPEAL," 1992

With a historical perspective of one hundred years from publication of *The New Atlantis* and seventy-five years from the founding of the Royal Society in England, Jonathan Swift made it abundantly clear that the new, specialized, and subsidized scientist of the Baconian cast had shown himself to be something less that a disinterested servant of humanity. In his description of the fictional Academy of Logado in *Gulliver's Travels*, he satirized the professional scientist who labored for pay in the new research institutions.

Swift's imaginary Academy had five hundred rooms in which scientists were pursuing all kinds of endeavors, including one to breed sheep without wool, another which had employed fifty men for thirty years to fashion pillows from marble, another to build a house from the roof down, and a fourth to prepare a treatise on the malleability of fire. The Academy also housed a huge computer-type machine, which its inventor claimed would enable even the most ignorant person to write books on philosophy, poetry, theology, etc., although the results came out scrambled and had to be pieced together by young students hired for the task. Despite the obvious shortcomings of the device, its inventor was portrayed by Swift as urging the public to raise a fund of money so it could be perfected.[68]

Gulliver described the first scientist he encountered in the Academy, a man who had worked eight years on a project to extract sunbeams from cucumbers, as meager, unwashed, and clothed in a drab manner. He also commented on the practice of all the research fellows of begging for money.[69] These amazingly prescient descriptions easily fit the behavior of contemporary research scientists at universities and private think tanks who spend a significant amount of time and energy politicking and conniving to secure governmental or corporate funding.

With the institutionalization and subsidizing of research by secular powers, the nature of scientific truth risked becoming whatever best served the career interests of the scientists involved. To provide a contemporary example, when the United Nations held an Earth Summit on the global environment in Rio de Janeiro in 1992, two hundred eighteen scientists, including twenty-six Nobel Prize winners, signed a petition ("The "Heidelberg Appeal"), which was conspicuously published on the op-ed page of the *Wall Street Journal*. Warning of the "false gods of Rio" (i.e., environmentalist concerns), the petition stated that "A natural state . . . probably never existed since man's first appearance in the biosphere, insofar as humanity has always progressed by increasingly harnessing Nature to its needs and not the reverse."[70]

This petition was evidence of the acute sensitivity of many scientists and technologists to the appearance of anything in public discourse that they perceive as threatening the principle of science and technology without limits. Its implication is that nature as such has no legitimate claim to be respected vis-à-vis science. The signatories of the Heidelberg Appeal demanded that environmental "stock-taking, monitoring, and preservation be founded on scientific criteria"—even though such criteria are quantitative, reductionist, and "objective" in the sense that they exclude any emotional or moral connection to nature when the scientist is professionally engaged in determining its "truth." In other words, these signatory scientists wanted the fate of the environment to be decided by their own, single-minded truth criteria, intended primarily for the mastery, rather than the sustainability, of nature. They illustrated their one-dimensional and unmitigated positivism by reducing the issue to a false dichotomy, stating in the Appeal that the "greatest evils which stalk our Earth are ignorance and oppression, and not Science, Technology, and Industry."[71]

Given the willingness of many scientists to provide the theoretical basis for technologies that have an alarming magnitude of potential harm to the environment, their ready dedication to military research, their financial subservience to vested governmental and corporate interests, their willingness to use flawed "science" to validate misleading claims by these interests, their too-frequent specialization and lack of any broader philosophical perspective, and their deliberate suppression of love and respect for nature in arriving at their professional judgments, why should their pronouncements monopolize the debate about the "truth" of nature or ultimately dictate the fate of the environment? Leaving everything in their hands would only accelerate the process of the destruction of nature in which we are already engaged.

The Recurrence of the Technology-Related Christian Themes

THE LIKENESS-TO-GOD AND IMPROVEMENT-OF-CREATION IDEAS

Since the Renaissance was a period of rebellion against prevailing ideas in science and philosophy, it was not surprising that some of the most important thinkers of that period described the new scientific knowledge and the technology it made possible as the acquisition of godlike power. This divine analogy had its origins in the pagan Prometheus myth, according to which technology was a theft from the gods, but it was also implicit in the Christian idea (Genesis 1:26) that humans were made in God's likeness.

In the fifteenth century, Marsilio Ficino, the Florentine Neoplatonist, considered the arts, which included technology, to be an indication of human immortality and likeness to God, despite the existence of original sin. Furthermore, he articulated the Christian improvement-of-nature theme: Not only did humans imitate all of the works of the divine nature, he said, but they even improved and perfected the works of nature.[72]

Ficino's contemporaries, Agrippa and Pico della Mirandola, were empiricists who practiced alchemy and wanted to develop a kind of natural magic that would give humans a godlike, Faustian power over nature.[73] Thus they, too, were dreaming of the kind of awesome, scientifically derived technological power that humans today, for better or worse, possess.

There was an element of biblical species-arrogance in Paracelsus's assertions in the sixteenth century about human godlike qualities and the mastery of nature. "Human nature," he wrote, "is different from all other animal nature. It is endowed with divine wisdom, endowed with divine arts. Thus, we are justly called gods and the children of the Supreme Being."[74]

In the view of Paracelsus, who was the most famous alchemist of the sixteenth century, humans revealed themselves in their works, just as God did. By engaging in a continual investigation of nature, he said, humans were discovering what God's gifts to them were.[75] Founder of the practical medical science of pharmacology, he saw technological activities as a praxis by which humans assisted God and revealed their being, including its divine aspects.

Humans also had an obligation to God, in his view, to improve what had been given to them. Thus, like Ficino, he brought up the Christian theme of finishing and improving God's creation, which had been articulated centuries before by patristic theologians, reiterated in Aquinas's idea of a "second perfection" of Creation by humans, and reaffirmed in John Calvin's pronouncement in the sixteenth century that nature could—and should—be improved by human action. Paracelsus stated that God had created nothing to perfection and had bid Vulcan, the master of fire and metalworking, to finish the process.[76] In other words, humans who discovered the secrets of nature and applied them to improve it were engaging in an exalted task.

Descartes's new subjective philosophy, which he promised would enable humans to accede to a godlike "mastery and possession" of nature, revealed a number of resonances with the Christian likeness-to-God theme. When he claimed that a human mind like his own could determine that the true reality of nature was mathematical and then discover its laws, one could say that he elevated it to the level of God's. Not surprisingly, Descartes said that a divine messenger, the Angel of Truth, had appeared to him in a dream in 1619 and had told him that mathematical thinking would "unlock the secrets of nature."[77] And perhaps most revealingly, when he wrote of his "unbounded will" in the *Meditations* (1641), he stated: "I bear the image and similitude of God."[78]

THE TECHNOLOGICAL-ADVANCEMENT-AS-RECOVERY-FROM-THE-FALL THEME

As evidenced by the persistence of such Christian themes in the face of the pagan revival which defined the Renaissance, it was obvious there would be no reversion at that time to an Apollonian theory of technology of limits. The idea of technological advancement as a means of recovery from the Fall, which was not conducive to the idea of limits, was voiced by both Paracelsus and Bacon.

From his Christian perspective, Paracelsus characterized the search for "natural magic" as a way to recover secrets lost because of the Fall. Echoing

Genesis, he stated that after the Fall, women had to bear children in sorrow and men had to work with their hands. When humans were expelled from Eden, he added, they had received some of the same knowledge as the angels. Thereafter, however, they had to ferret out the secrets of nature on their own by means of craft. Humans thus had an obligation, he concluded, to investigate nature and discover the arts, including alchemy, which had not been given to them by God in immediately recognizable form.[79]

The most important promoter of the Christian recovery theme in the seventeenth century was Bacon, who despite his advocacy of a new method in science, made it clear that he was interested in technological applications. The British Puritans, whose millenarian perspective he shared, believed that education would restore the mind to its prelapsarian powers, that medicine would facilitate the recapture of lost physical perfection, and that science and technology would restore lost human control over nature.[80] Bacon, therefore, saw the mechanical arts and the sciences as a crucial element in recovery from the Fall. He mentioned that when Adam and Eve were expelled from the Garden of Eden, God had said, "[I]n the sweat of thy brow shalt thou eat bread."[81] This meant for him that hard work was essential for recovery; but it had to be combined with the "proper method" of a new science that could provide technological power over nature.

Bacon contrasted the stagnation of the existing sciences with the dynamism of the mechanical arts (i.e., technology), which he believed had an inherent developmental tendency and were therefore "perpetually improved."[82] They "are founded on nature and the light of experience," he said, ". . . and uninterruptedly thrive and grow."[83] Little did Bacon realize at the time he wrote those words, however, how prophetic they would be. Today, the never-ending process of technological development has even become for some ideologues and technologists an end in itself, regardless of any negative effects or alarming consequences.

Bacon, in any case, wanted a new scientific method that would yield knowledge that would make it possible to extend and accelerate the process of technological development. He made clear in *Novum Organum* that technology was uppermost in his thinking when he cited the compass, gunpowder, and the printing press as examples of practical inventions which, in contrast to the sterile mental artifacts of Scholastic natural philosophy, were changing the world.

Moreover, in his idealized vision of a scientific research center, Solomon's House, Bacon pointedly spoke of the benefit of a number of technological advances. "We represent," he wrote of the promised discoveries of his researchers, "also ordnance and instruments of war, and engines of all kinds, and likewise new mixtures and compositions of gunpowder, wildfires burning in water and unquenchable." It was obvious from such claims that he was eager to win over his readers to his new science; but as it turned out, some of the technological innovations he pictured were ultimately achieved without any direct assistance from science. [84]

Some of his descriptions had similarities with the visions of his thirteenth-century predecessor, Roger Bacon. "We imitate also flights of birds," he said, adding that "we have some degrees of flying in the air; we have ships and boats for going under water, and brooking of seas; also swimming girdles and supporters."[85]

These ambitious dreams prefigured the technological development in the twentieth century of aircraft, submarines, and scuba-diving equipment. They became realities that are now part of our vast repertory of technology without limits, whose scope and power actually extends far beyond the innovations that Bacon had imagined. The emergence of instrumental science in the seventeenth century, along with Bacon's new methodological "tool" for the mind and the new standard of practical applicability implicit in Bacon's dictum that knowledge equaled power, was another reason for a new, dynamic relation between science and technology. Not surprisingly, Bacon made it clear he understood the important contribution improved technological instruments would make to the advance of scientific knowledge. For example, he expressed an interest in developing devices that would further the scientific investigation of outer space. "We procure means of seeing objects afar off," he wrote, "as in the heavens and remote places . . ."[86]

ANTHROPOCENTRICISM: LEGITIMIZATION OF SCIENTIFIC EXPERIMENTS ON ANIMALS

Given his aggressive stance vis-à-vis nature and a preoccupation with such Christian themes as recovery from the Fall and the perspective of a millennium, it was in no way surprising that Bacon approved of the use of animals for scientific research. Of his imaginary research center, Solomon's House, Bacon wrote:

> We also have parks and inclosures of all sorts of beasts and birds, which we use not only for view or rareness but likewise for dissections and trials, that thereby we may take light what may be wrought upon the body of man.[87]

In other words, animals would be used in experiments to discover the effects of substances and procedures, which would shed light on whether they might be dangerous or harmful for humans. In addition, Bacon described breeding experiments that prefigured today's transgenic manipulations in biotechnology. He said: "We also make them [the animals] differ in colour, shape, activity, many ways. We find means to make commixtures and copulations of different kinds, which have produced many new kinds . . ."[88]

Both types of experiment meant the objectification of the animals used in research and the absence of concern for any limits on their legitimate use by humans. Furthermore, Bacon's anthropocentrism was so thoroughgoing that he implied that such breeding experiments did not have to satisfy an immediate utilitarian goal. They could be done for their own sake—presumably to satisfy curiosity or to affirm the brilliance of the scientist. Ironically, or perhaps fittingly, Bacon died of a respiratory illness contracted during a snowstorm while he was experimenting with freezing a hen.

Bacon's unloving attitude toward animals was consistent with the anthropocentrism of the Judeo-Christian nature ethic, which was also reflected in Descartes's denial of a reasoning power to animals. According to Descartes, the bodies of humans and animals were nothing more than machines, like a clock or some other automatic device. He believed, furthermore, they were capable of functioning without a soul, although humans had a rational soul and animals did not. The proof for Descartes that animals could not reason was his belief that unlike humans, they had neither language nor the capability to adapt to new situations.[89] By thus emptying them of any *cogito*, he reduced animals to the level of things to be dominated. For him, they belonged to the objectified nature over which the human subject was to be the master and possessor. And if they possessed no soul or power of reason, then why should they be worthy of love or respect?

Not all thinkers of the period agreed with Descartes on this point, although his view was consistent with Aristotle's assertion that it was reasoning ability "more than anything else which is man."[90] In contrast to Descartes, his sixteenth-century predecessor, Montaigne, had written of the "discourses of beasts;" and in an apparent reference to elephants' behavior when one of their kind was dead or dying, he stated that they "had some apprehension of religion."[91] He also maintained that the existence of animal language and the extraordinary activities of animals like swallows, bees, and ants were evidence that animals possessed reason. Furthermore, the distance in intelligence between humans and animals, he argued, was sometimes less than that which existed between individual humans.[92] Gassendi, a contemporary of Descartes in France, maintained that animals were only capable, however, of what he called "sensitive reasoning."[93]

EMPHASIS ON HUMAN AGENCY AS A STEP TOWARD SECULARIZATION OF THE CHRISTIAN VIEW OF HISTORY

In terms of the eventual emergence of a culture of technology without limits, the linkage of the new Baconian power knowledge with the Christian theory that history had a linear direction toward a positive goal was one of the most important ideological developments of the Renaissance. It was a history, however, that was in the process of being desacralized and that would be fully transformed in the Enlightenment. Machiavelli, although he recognized a role of fate in historical outcomes and saw history running in cycles, emphasized in *The Prince* the importance of human agency in historical outcomes. Similarly, Jacques-Bénigne Bossuet, who adhered to the Christian view that God determined history in his *Discourse on Universal History* (1681), left room for cyclic patterns and free play on some occasions for the effect of human passions, wisdom, and ignorance.[94]

More influential in its long-term effect, however, was the biblical idea of the millennium, which included both human agency and a divine factor and had been resurrected in the late Middle Ages by Joachim of Fiore and some of the radical Franciscans. In the sixteenth and seventeenth centuries, it attracted new adherents among Christians and Jews. In this later version, it encouraged the belief that technological advances leading to improvement of the human condition were necessary for the realization of a divine plan. According to the theory, the coming of a messiah and the inauguration of a thousand-year period of betterment would not actually occur *until* humans themselves had begun a process of improvement and recovery from the Fall.[95]

After the expulsion of Jews from Spain at the end of the fifteenth century, some mystical adherents to the new Lurianic sect of the Jewish Cabala believed a messiah would appear when the Jews, on their own, began to recapture the goodness of Adam and Eve before the Fall. And in Germany, members of the Rosicrucian order had a similar belief that there would be a new dawn of human power and knowledge followed by the end.[96]

Although St. Augustine had interpreted the Book of Revelation, which predicted a Second Coming and millennium, as having only allegorical meaning, and although Martin Luther in the sixteenth century had theorized that the millennium had already passed and that the end of the world was near, the idea of an approaching millennium caught on in seventeenth-century England among some Christian thinkers. Dr. Joseph Mede of Cambridge, for example, argued that the course of history since Christ was one of gradual defeat of evil and that human betterment signaled the coming of the millennium.[97]

Henry Archer was another who believed there would be an increase of human knowledge before the Second Coming; and he described the approaching millennium as a temporal (i.e., earthly) state of peace, safety, riches, health, and long life. The great Puritan poet John Milton declared in *Aeropagitica* (1644) that a millennium was near and that it would be worldwide.[98]

Francis Bacon's millenarian views reflected those of the British Puritans, and his *New Atlantis* (1627) revealed as well the influence of Rosicrucians and

Lurianic Cabalists. He believed humans could prepare the way and even hasten the reappearance of Christ and the inauguration of the millennium by achieving the "Great Instauration" of learning, which his new scientific method was designed to promote. Humans would thus recapture through their own efforts the knowledge and power over nature that Adam enjoyed before the Fall, a necessary step before the millennium could follow.[99]

This linking of the Christian recovery theme with the idea of the millennium placed considerable emphasis on the idea of human agency, although it was only in the secular theory of progress that humans would be seen as responsible for the whole process of betterment, from start to finish. Bacon's claim that his new science would help inaugurate the millennium implied, in any case, that scientific and technological advances would inevitably benefit humanity. Perhaps this is why he did not pay more attention to the question of the moral responsibility of the scientist.

In France, Descartes anticipated the modern, secular theory of progress when he presented in his *Discourse* (1637) a new philosophical method that would enable humans to act upon the world independently of God and render themselves "masters and possessors of nature." He did not go so far as to include the idea of continual and inevitable betterment over time, however; but he clearly had in mind a collective project by which humans would expand their power over nature through technological applications of a new science.

According to Hannah Arendt, Pascal was the European thinker who, in his *Preface to a Treatise on the Vacuum* (1647), first saw the human race as the subject of a never-ending, progressive cultural evolution over time.[100] In contrast to animals, said Pascal, "whom nature maintained in a limited order of perfection ... the human is only made for infinity ... [H]e educates himself continually as he progresses."[101] It was the "particularly [human] prerogative," Pascal added,

> ... that not only each human being can daily advance in knowledge, but that all men together progress continually while the universe grows older ... so that the whole succession of men throughout the centuries should be considered as one and the same man who lives forever and continually learns.[102]

It is important to note, however, that because of his strong Christian commitment to the idea of a fallen human nature, Pascal emphasized progress in knowledge rather than moral improvement. He adhered unequivocally in his *Pensées* (1670) to Augustine's negative view of human nature, hopelessly depraved after the Fall and only capable of salvation by the grace of a God upon whose uncertain existence he, Pascal, urged humans to "wager."

Notes to Chapter 3

1. J. S. Spink, *French Free-Thought, from Gassendi to Voltaire* (The University of London: The Athlone Press, 1960), 87, 114.

2. Renè Descartes, *Discours de la Méthode*, ed. M. Robert Derathe (Paris: Librairie Hachette, 1937), 48, 53, 56.

3. Spink, 89.

4. Isaac Newton, *Optiks* (New York: Dover Publications, 1952), 400, 403.

5. Alexandre Koyrè, *From the Closed World to the Infinite Universe* (Baltimore: Johns Hopkins Press, 1957), 6-8.

6. Ibid., 31.

7. Ibid., 39.

8. Blaise Pascal, *Pensées* (Paris: Hachette, 1950), 41, 89.

9. Koyrè, 187-88.

10. Hiram Hayden, *The Counter Renaissance* (New York: Charles Scribner's Sons, 1950), 114.

11. Arendt, Part 2, 26.

12. Dyson, 13.

13. Pascal, 63, 65, 68.

14. Ibid., 38-41.

15. William Leiss, *The Domination of Nature* (New York: George Brasiller, 1972), 151-52.

16. Glacken, 473.

17. Descartes, 69.

18. John Zerzan in *Fifth Estate* 20, no. 2 (1985): 26.

19. Thomas Goldstein, *Dawn of Modern Science* (Boston: Houghton Mifflin, 1980), 127.

20. Hayden, 338.

21. Koyré, 132.

22. Neil Van Deusen, *Telesio: The First of the Moderns* (New York: Doctoral Thesis, Columbia University, 1932), 19, 45.

23. Max Horkheimer and Theodor Adorno, *Dialectic of Enlightenment* (New York: Continuum Publishing Company, 1972), 3-4.

24. Hayden, 201-03.

25. Niccolo Machiavelli, *The Prince*, in *Masterworks of Government*, ed. Leonard Dalton Abbott (New York: McGraw Hill Book Company, Masterworks Series, 1973), vol. 1, 162.

26. Van Deusen, 25, 89.

27. Hayden, 255, 266.

28. Francis Bacon, *Essays, Advancement of Learning, New Atlantis, and Other Pieces*, ed. Richard Foster Jones (New York: Odyssey Press, 1937), 274.

29. Ibid., 272.

30. Ibid., 315.

31. Hayden, 264.

32. Ibid., 264-5.

33. Ibid., 263.

34. Descartes, 69.

35. Ibid., 50 n.3, 51n. 1.

36. Ibid., 47.

37. Ibid., 48, 53, 56.

38. Ibid., 69.

39. Michael Zimmerman, "Current Debate: Nature and Domination," *Tikkun* 4, no. 2 (1989): 103.

40. Descartes, 70-71 n. 1, 71.

41. Martin Heidegger, "Age of the World Picture," in *The Question Concerning Technology and Other Essays* (New York, Harper Torchbacks. 1977), 127.

42. Ibid.. 130-31.

43. Ibid., 133-34.

44. Butterfield, 112.

45. Hayden, 293-94.

46. Glacken, 471.

47. Van Deusen, 41.

48. Miriam M. Reik, *The Golden Lands of Thomas Hobbes* (Detroit: Wayne University Press, 1977), 73.

49. E. A. Burtt, *The Metaphysical Foundations of Modern Science* (Garden City, N.Y.: Doubleday Anchor Books, 1954), 226.

50. Theodore Roszak, *Where the Wasteland Ends* (New York: Anchor Books, 1973), 146.

51. Ibid., 150.

52. Bacon, 279.

53. Ibid, 283.

54. Friedrich Nietzsche, *The Will to Power*, ed. Walter Kaufman (New York: Vintage Books, 1958), 50.

55. Roszak, 143.

56. Glacken, 472-73.

57. Bacon, 284.

58. Ibid., 329.

59. Butterfield, 116.

60. Barry Commoner, *The Closing Circle* (New York: Bantam Books, 1972), 185.

61. Butterfield, 115.

62. Ibid., 119.

63. Bacon, 276.

64. Jacques Ellul, *The Technological Society* (New York: Vintage Books, 1964), 85-86.

65. Notes taken by the author from a Winter, 1982, lecture by Professor Roger Hahn at the University of California at Berkeley.

66. Bacon, 488.

67. Ibid., 489

68. Jonathan Swift, *Gulliver's Travels* (New York: Rand McNally and Company), 201-05.

69. Ibid., 201.

70. *Wall Street Journal*, 1 June 1992.

71. Ibid.

72. Glacken, 463.

73. Hayden, 179, 183-84.

74. Noble, 36.

75. Glacken, 466-67.

76. Ibid.

77. Burtt, 105.

78. Arendt, Part 2, 26.

79. Glacken, 466-67.

80. Francis Yates, "Science, Salvation, and the Cabala," *New York Review of Books,* 27 May 1976, 27.

81. Glacken, 472.

82. Ibid., 474.

83. Ibid.

84. Bacon, 487.

85. Ibid., 488.

86. Ibid., 486.

87. Ibid., 483.

88. Ibid.

89. Descartes, 63-64 n. 4.

90. Aristotle, 265.

91. Hayden, 399-400.

92. Descartes, 65 n. 2.

93. Spink, 99.

94. Andre Lagarde and Laurent Michard, ed., *XVII siècle: Les Grands Auteurs Français de Programme III* (Paris: Editions Bordas, 1970), 280-81.

95. Yates, 28.

96. Ibid.

97. Ernest L. Tuveson, *Millennium and Utopia* (Berkeley: University of California Press, 1949), 77-78.

98. Ibid., 86, 92.

99. Ibid., 27.

100. Arendt, Part 2, 153.

101. A. Chassang and Charles Senninges, ed., *Recueil de Textes Littéraires Françaises, XVII Siècle* (Paris: Hachette, 1975), 132.

102. Arendt, Part 2, 153.

4

THE ENLIGHTENMENT

The Implications of New, Eighteenth-Century Ideas about Nature

THE REMOVAL OF GOD FROM THE NEWTONIAN UNIVERSE

According to Newton's science, the universe was a vast mechanism of infinite temporal and spatial dimensions, designed and built by a clockmaker God, whose existence he inferred. Since it showed the logical necessity of God as a first cause and architect of this clockwork universe, Newton believed his *Principia* would strengthen belief in a Divine Being. He would have been shocked if he had lived to see that by the end of the eighteenth century, scientists who had accepted his model of the cosmos would modify it to remove the physical need for God as either a first—or, for that matter, final—cause. Such a modification would mean that from a scientific perspective, the created world of nature was, as had been the world of the pagan atomists, a world without higher purpose.

Early in the eighteenth century, the German philosopher Leibnitz openly declared his disagreement with Newton's theory. In the first place, Leibnitz's universal principle of sufficient reason did not allow for a vacuum in nature, which Newton had correctly argued existed. Second, Leibnitz asserted that God had made the best possible world, a kind of perpetual motion machine. This view conflicted with Newton's claim that God was a clockmaker who had to intervene periodically to rewind and clean the cosmic mechanism. Third, again in opposition to this idea of a largely absent clockmaker God, Leibnitz argued that Divine Providence controlled every single occurrence, both human and natural, that took place in the phenomenal world.

The publication in 1706 of the second edition of Newton's *Opticks*, which contained his statements about God, was the catalyst that led to the famous debate between Newton and Leibnitz concerning God and the validity of Newton's world system. Leibnitz accused Newton of introducing "occult qualities" into natural philosophy with his theory of gravity, whose cause Newton said he did not know but which he suggested was spiritual. Newton responded to this charge in his *General Scholium* (1713), where he stated that although he had not been able to determine the cause of gravity, "I frame no hypotheses; for whatever is not determined from phenomena is to be called hypothesis; and hypotheses, whether metaphysical or physical . . . have no place in experimental philosophy."[1]

Furthermore, Newton conceded, in apparent agreement with Leibnitz's criticisms, Divine Providence was more than a largely absent clockmaker God. He was an active force in the world, said Newton, the God of the Bible—not only a First Cause, but also the God of final causes. Newton acknowledged, however, that "a god without dominion, providence, and final causes, is nothing else but Fate and Nature."[2] There was a certain irony in these words, however, since they tacitly conceded that if his explanation of the universe did not need a providential God to complete it, it would amount to a purely secular—and purposeless—view of nature.

In fact, despite Leibnitz's preoccupation with the question of the nature of God and his role in the world, the questions he raised led nearly a century later to the purely secular Newtonian physics that both atheists and deists wanted.[3] At the end of the eighteenth century, the Frenchman Laplace supplied exclusively scientific explanations for the divine elements in Newton's theory. In 1786, he demonstrated mathematically the stability of the solar system at its outer reaches and thereby eliminated the need for a God who returned from time to time to regulate and rewind the mechanism of the universe.[4] Without this need, the idea of providence was superfluous to Newtonian science, and God was not even the Final Cause Newton had suggested he was.

Laplace also undermined the idea that God was the First Cause of the universe with his Nebular Hypothesis of 1796, a mathematical demonstration that it was possible to explain its origins in terms of purely physical mechanisms. In 1802, Laplace reportedly had a conversation about his findings with the French emperor, Napoleon, during which he stated, in an obvious rebuke to Newton, that God himself was nothing more for science than a "hypothesis."[5]

The implications for science, religion, and philosophy of this secularization of the Newtonian model of the universe were, indeed, profound. The seventeenth-century compromise worked out by thinkers like Descartes, Gassendi, and Newton between the atheistic science of the pagan atomists and theological explanations based on the God of the Bible was effectively shattered. For science, the universe was henceforth only a vast, impersonal mechanism whose true nature was mathematical. It was governed not by divine will, but by laws divorced from any higher purpose. The post-Apollonian project to master nature, essentially compatible with the anthropocentric nature ethic of Genesis, had been formulated almost a century before Laplace by thinkers like Bacon and Descartes. His mathematical demonstrations of a secular and nonteleological world of nature, however, provided an incentive for mastery that was based exclusively on considerations of power and security rather than on human assumption of a divinely assigned role to be the master and ruler of nature.

And in the face of doubts about whether humans had any special importance in the new, mechanistic universe of the scientists, a number of the leading Enlightenment *philosophes* clearly opted for humanism, not fatalism or despair. Kant, for example, said that humans were the final end of nature, since they were the only element in it capable of freely choosing action in opposition to their natural inclinations.[6] He thus demonstrated his allegiance to a humanistic strain in the Western cultural tradition that could be traced back to classical Greek civilization. Diderot was another Enlightenment *philosophe* who affirmed the values of humanism, specifically in response to the cold and mechanistic view (with or without a God) of an infinite universe in which humans were an insignificant presence. Without humans, he said, "this moving and sublime spectacle of nature would be nothing more than a sad and mute scene ... It is the presence of man which makes the existence of beings meaningful."[7]

LITERARY PROMOTION OF ROMANTIC AND SADISTIC HUMAN-NATURE RELATIONS

Rousseau's exaltation of sentimental love for nature. At a time when scientists had mathematized nature and insisted on objectivity to distance themselves emotionally from nature in order to attain mastery over it, Rousseau and the Marquis de Sade expressed in literature new possibilities, one positive and the other negative, for an erotic relation of humans to nature that might have modified their actions toward it. In the case of Rousseau, he wrote extensively and lyrically of his love of nature. Whereas the Irish philosopher Bishop Berkeley argued that what humans experienced as nature was only a product of the mind, and Kant said that the mind "constructed" the world of nature from the sensuous manifold of external phenomena by bringing such concepts as space, time, and causation to it, Rousseau treated nature as a rich and independent sensuous entity that inspired wonder and positive emotion on the part of humans. He described the joys of a close human relation to nature in such works as *The New Héloïse* (1761) and *Reveries of a Solitary Wanderer* (1782).

The important question is, however, did his attitude and that of the other Romantic artists and thinkers of the late eighteenth and early nineteenth centuries

represent a significant opening in European intellectual culture for a loving re-
conciliation of humans with nature rather than a preoccupation with mastery?
Unfortunately, the answer cannot be given in the affirmative in Rousseau's case
because he did not express an authentic love for nature in itself. Despite all his
gushing declarations of love for it, in truth, nature only had value for him as a
means to satisfy his own particular needs—that is, to provide refuge from a
corrupt society and to function as a muse which, in the absence of human contact, stimulated his senses and feelings and inspired his thoughts, daydreams,
and literary creations.

The problem was that although Rousseau was a romantic who wrote elo-
quently about feelings and emotions and said humans were naturally good, he
expressed antipathy for most of his human contemporaries and a sense of
isolation and estrangement. In effect, he had a narcissistic temperament and a
paranoid sensibility, wounded by years of criticism by other *philosophes* who
disagreed with his controversial views on such matters as progress, education,
religion, and civilization and by what he felt was a general rejection of his ideas
by society at large. In his *Letter to D'Alembert on the Theater* (1757), he clearly
revealed his own marginality when he identified himself with Alceste, the
solitary misfit of Molière's most famous play, *The Misanthrope*.[8] In his *Reveries*
(1782), written in the closing years of his life, Rousseau complained bitterly:
". . . [M]y fellow men . . . are strangers, unknown, nothings because that's what
they wanted."[9]

Typically for Rousseau, his love of nature was always in opposition to his
disdain for human society. It provided him with an emotional vehicle for
transcending the shortcomings of its members and healing his wounded and
isolated self. Thus, it was in contrast to this humanity, which he rejected and
which he felt had rejected him, that he wrote in *The New Héloïse* of the uplifting
effect of being in nature, in the Alps, away from the rest of humanity: "It seems
that in rising above the areas inhabited by humans, one leaves behind their low
and terrestrial sentiments . . ."[10] Similarly, in his *Rêveries*, another exculpatory
work produced in his waning years, Rousseau described his solitary visits to the
lake island of St. Pierre in Switzerland and explained how they provided him
with compensations for the happiness denied him in human society. It was a
place, he wrote, where one could "intoxicate himself, at his own leisure, with the
charms of nature." There, he said, a person "sufficed to himself, like God." He
could live totally in the present, abandoned to himself, to nature, and to his
thoughts.[11]

Sade's option for cruelty and destruction of nature. The active and deviant
counterpoint to Rousseau's passive, sentimental love for nature as muse and
refuge was the perverted eros of the Marquis de Sade. For him, nature was not
the benevolent and protective presence it was for Rousseau. Rather, it was a
dark and evil force that replaced God as the overwhelming and arbitrary power
of the world. "Cruelty is natural," Sade declared in *Philosophy in the Bedroom*.

Benevolence and "what fools call humanness," he wrote, "have nothing to do with nature" and are "the fruit of civilization and fear." Thus, in response to the question, what is the difference between humans and animals and plants, Sade responded, "None, obviously."[12]

By urging rebellious humans to identify with the immorality of nature in their sexual behavior, Sade demonstrated an affinity with the erotic doctrine of those seventeenth-century French libertines who had asserted that one's erotic actions should reflect one's particular nature. Sade's ideas about sexuality, however, were much more extreme. He promoted cruelty, sodomy, torture, and a complete dehumanization of the love object. And just as the scientist distances himself from what he seeks to control by means of objectification and quantification, his character Clairwill boasted in his novel *Juliette* of her "serene command over her emotions," which permitted her "to do ... everything without any feeling."[13]

In his writings, Sade advocated depersonalized and quantified group sex. He described elaborate pyramids of individuals engaged in various forms of copulation. In essence, he did not recognize any limits in the conduct of sexual relations. It should be recognized, however, that such ideas were the product of the imagination of a man who spent twenty-nine years of his adult life in prisons and a mental institution. As Camus said of Sade, "In the depths of the prison, dreams are without limits and reality interferes with nothing."[14]

Our interest in Sade derives precisely from the relevance of his professed desire to torture and destroy what he was unable to love to today's destruction of nature by technological civilization. Significantly, Sade's destructive urge and nihilism were such that he even extended his cruelty to external nature. "I abhor nature ...," Sade declared in a remarkable and revealing passage. He added:

> I should like to upset its plans, to thwart its progress, to halt the stars in their courses ... to destroy what serves nature and to succor all that harms it ... We could perhaps attack the sun ... or use it to set fire to the world ...[15]

This was the declaration of a sadist, ready to destroy that which he should love but is incapable of loving. Today, one is tempted to conclude that a similar perverted urge drives a massive and destructive technological assault on nature. In terms of instinctual dynamics, it could be said that today's culture of technology without limits appears to have the trappings of a culture of technological sadism directed at nature.

THE CHALLENGE OF BUFFON'S THEORY OF EVOLUTION TO THE BIBLICAL CREATION STORY

The theory in Genesis of a divine creation of separate and immutable natural species in the beginning, including humans whom God uniquely created in his own image, was challenged in the eighteenth century by Georges-Louis Buffon's dynamic, pre-Darwinian theory of evolution. According to Buffon's *Natural History* (1751), the days of the Creation mentioned in Genesis were really only epochs, humans were just another animal species, and nature represented an ongoing process of destruction and renewal rather than a fixed repertoire of living species created once and forever.[16] When Darwin reiterated these ideas in the nineteenth century, their subversive impact on Christian belief helped pave the way for nihilism.

After members of the Faculty of Theology at the Sorbonne University in Paris complained that Buffon's writings cast doubt on the Bible, however, he publicly stated: "I give up everything in my book . . . that could be contrary to Moses' narration." In private, however, he held to his view that humans had an animal nature. He also privately maintained that the age of the earth could be scientifically derived from empirical facts.[17] This belief challenged, as did writings of other Enlightenment *philosophes* like Diderot, Holbach, Hume, and Kant, Archbishop Ussher's widely accepted, scripturally based estimate at mid-eighteenth century that the Creation described in Genesis had occurred at a precise year, 4004 B.C.[18]

THE IDEA OF MALLEABLE HUMAN NATURE AS CONTRADICTING THE DOCTRINE OF ORIGINAL SIN

Rousseau broke with the self-depreciating Christian view of the self, irremediably besmirched by original sin. In his first and second *Discourses* (1751 and 1755), he argued that humans were naturally good. It was organized society and the advance of civilization, he said, that had made them bad. Humanity had been corrupted by the arts and sciences, the education system, government by the rich, and the institution of private property—all the products of civilization. Thus, in his first *Discourse*, Rousseau compared the happiness and morality of tribal peoples living close to nature with those of Europeans and found the former unquestionably superior.

Some *philosophes* redefined human nature in a way that eventually invited the application of technology to humans for the purpose of manipulation and control, although it is unlikely that they, themselves, would ever have approved of such a deliberate use. For example, both Locke and Helvetius saw human nature as a blank slate, but their goal was to mold it for the better by means of laws and education that essentially respected individual autonomy. Voltaire adopted a similar stance in his *Philosophical Letters* (1734). He found Pascal's essentially pessimistic, Augustinian view of human nature excessive and argued that humans were both good and bad. For his part, Diderot stated it was society, not human nature, that was responsible for evil. "It is bad education, bad models, bad legislation," he said, "that corrupt us."[19]

The historical theory of progress, of which Antoine de Condorcet was one of the leading proponents, included the idea of malleable human nature. For example, he argued that ". . . nature has set no terms to the perfection of human facilities" and believed that education could rectify even the differences in individuals' native abilities.[20] He even advanced the view that acquired characteristics could be genetically transmitted. Education, he said, would "modify the physical organization" that determined the power of the brain, and individuals would then pass such changes on to their offspring.[21] In the twentieth century, however, the idea of a malleable human nature lent legitimacy to the claims of behavior control technologists like B. F. Skinner, who advocated engineering the behavior of humans to make them better.

Other Challenges to the Established Basis of Higher Values

THE *PHILOSOPHES'* CRITIQUE OF RELIGION AND SKEPTICISM ABOUT GOD

Although Bacon had situated his appeal for the development of a new science of technological application in the context of the Christian idea of re-covery, a subsequent decline of Christian belief nevertheless had the effect of preparing the philosophical ground for the emergence of technology without limits. It did so by contributing to the collapse of the entire framework of higher values that was necessary to orient technological development and control technology's use.

In the eighteenth-century Age of Reason, both organized religion and the God postulate were subjected to more incisive critical scrutiny. Voltaire, for example, had nothing but scorn for organized religion. He said the Christian apostles were "simple, credulous idiots" and characterized the church fathers as "charlatans and deceivers."[22] In his *Philosophical Dictionary* of 1764, he described both the Old and New Testaments as full of absurdities and irreconcilable contradictions.[23] Nevertheless, Voltaire's famous remark, "If God did not exist, we would have to invent him," demonstrated he understood the practical value of human belief in a divine being. "I want," Voltaire said, "my attorney, my tailor, my servants, and even my wife, to believe in God, and I think that I shall be robbed and cuckolded less often."[24]

Diderot addressed his letters to Voltaire, "Dear Antichrist," after Voltaire openly called for the elimination of the influence of the Church of Rome. For his part, Diderot not only rejected Christianity and became a deist, but he eventually embraced atheism. He stated that Christianity recruited its monks and nuns from those who were sick in spirit,[25] and he remarked that the God of the Old Testament "was more concerned about his apples than his children."[26] In 1759, after Diderot included in his *Encyclopedia* an article written by D'Alembert that complimented some Protestant clergy in Geneva for believing Christ was not divine, the French government withdrew its royal privilege from that work.[27]

Diderot saw religion, in any case, as an alienation from the phenomenal world. "In vain," he wrote, "O slave of superstition, do you seek your happiness beyond the limits of the world in which I have placed you . . . Return to nature, to humanity, and to yourself again."[28] Furthermore, Diderot saw sin as deriving from Christianity itself, not human nature. Writing of the effects of Christian civilization on the indigenous peoples of the South Sea islands, he warned them: "One day they [Europeans] will come, with crucifix in one hand and dagger in the other . . . one day under their rule you will be almost as unhappy as they are."[29]

Eighteenth-century deists in Germany like Leibnitz, who based his argument on the principle of sufficient reason, tried to find a rational, nonrevealed basis for the existence of God. Deists generally believed in a supreme being who did not answer prayers, in an afterlife, and in an immortal soul.[30] They rejected miracles, priests, and the idea that human nature was depraved.[31] Eventually, however, they fell back on subjective justifications, which depended only on the individual, for the existence of God. In France, Rousseau wrote in *Emile* (1762) that the only way to know God existed was to feel his presence in nature. When he had been a deist, Diderot, too, argued that true religion was not based on revelation, but on what one felt.[32]

In his *Dialogues on Natural Religion* (1779), Hume demolished traditional rational arguments for the existence of God such as the argument from design and the theory that there had to be a Necessary Being to create the world. In response to the first, Hume argued that the fact that there were laws of phenomena and some measure of order in the universe did not necessarily imply that an intelligent being had designed it. And he pointed out that if there had to be a Necessary Being as Creator, since something supposedly could not derive from nothing, an infinite regression would result as one searched always for an earlier cause.[33]

Adopting a psychological approach that reflected his skepticism, Hume found the true roots of religion in emotions like hope, fear, and self-flattery on the part of the human species.[34] Religion, he said, was a palliative that made the misery of life bearable and provided a basis for morality. Like Voltaire, he concluded that religion was a necessary social institution. For practical reasons, he counseled humans to behave as if they believed God existed, even if in reality they did not believe he did.[35]

Kant's approach to the God question was similar. In his famous *Critique of Pure Reason*, he showed that pure reason (*Vernuft*) could not logically prove that a necessary being like God existed either inside or outside the phenomenal world. He concluded, however, that humans would be better off if they opted for a "rational faith" in his existence and behaved as if it were proven. Despite the practical advantages and his good intentions behind this "as if" approach to God, it was an important milestone on the way to the nihilism that emerged after Nietzsche proclaimed the "death of God" and the collapse of all higher values at the end of the nineteenth century.

THE *PHILOSOPHES'* CRITIQUE OF REASON IN THE AGE OF REASON

In contrast to the questioning of religion, a number of the most important thinkers of the eighteenth century expressed strong confidence in the power of human reason. Voltaire stated that there had been no human progress until the Age of Reason. Turgot argued in his famous 1750 lectures at the Sorbonne that as humans progressed, reason would slowly blot out emotion, until ignorance, prejudice, fanaticism, and superstition disappeared. In 1793, Condorcet saw a similar link between reason and human progress. Speaking of the future, he stated "men . . . will recognize no master but their reason."[35] As the Age of Reason gained maturity, however, it was inevitable that the validity of reason itself would eventually be subjected to the *philosophes'* critique, despite the fact that a considerable number of them had put their greatest faith in it.

Rousseau's advocacy of sentiment over reason. In the *Discourses*, written at mid-century, Rousseau launched his famous attack on reason. A man who cogitated, he declared in his first *Discourse*, was a "depraved animal."[37] In his second *Discourse*, he argued that education had been perverted by too much emphasis on the intellect. In opposition to reason, Rousseau glorified feeling and sentiment. And he stated, in contrast to his European contemporaries, that primitive peoples had been able to keep their happiness and virtue. The secret of virtue, he concluded, was to obey one's inner conscience.

Hume's rejection of reason and advocacy of skepticism. The "most radical of the radical philosophers" of the Enlightenment, according to Peter Gay, was Hume,[38] who showed the limits of both reason and empiricism. Whereas Rousseau argued that feelings, not reason, would be the proper guide to morality and happiness, Hume asserted that all cognitive activity was determined by feelings. "When I give preference," he said, "to one set of arguments over another, I do nothing but decide from my feeling . . ."[39] "Reason," he stated, "is and ought to be the slave of the passions and can never pretend to any other office than to serve and obey them."[40]

Hume argued, therefore, that what he called the imagination, stimulated by the passions, was the primary faculty of the human mind. He rejected the concept of a durable inner self, saying that there was actually only a series of continually changing ideas or impressions.[41] There was no thinking subject for him, therefore, in the Cartesian or Lockean sense. He psychologized the Cartesian ego,[42] in effect, by asserting that the most important philosophical concepts were based on psychological needs rather than on rationally determined truths. He said, for example, that the principle of causation was invented by humans to serve the need of providing a convenient and reassuring framework for their experience in the phenomenal world.

Habit and custom, which derived from psychological needs, were the real basis for Hume of moral standards and social institutions. These acquired a normative standing, he said, because they were the product of a long historical process. In other words, the way things had turned out to be over time was the way they should be. For Hume, therefore, history was the "mistress of wisdom." We should study it, he asserted, in order to understand the consistencies of human behavior and the motivations behind it.[43]

In Hume's view, science, like religion, was also not based on reason. The true nature of matter, he asserted, could not be accurately known by the human mind. Furthermore, he not only demonstrated that causation was based on habitual experience and could not be proven by reason alone, but this conclusion carried the implication that a pure empiricism, without rational induction, was also impossible. That is, because humans were obliged to make rational inferences from experience, they were unable to guarantee by either reason or experience alone the certainty of future recurrence upon which positive science depended.[44]

Hume's questioning of everything, according to Harry Prosch, had the effect of leaving humans "stranded with no firm conviction in anything."[45] Russell commented that "the growth of unreason" in the nineteenth and twentieth centuries was ". . . a natural sequel to Hume's destruction of empiricism."[46] Stanley Rosen concluded in his recent essay on nihilism that Hume's philosophy undermined the humanity itself of men and women by attacking the rationality upon which it was based. For him, Hume's radical skepticism marked a decisive step in the direction of contemporary nihilism.[47]

There was another respect in which Hume's thought had nihilistic implications. A mitigated skeptic, he recommended that humans should believe and reason as if God existed and as if causation were a fundamental characteristic of nature, although his philosophy claimed to demonstrate that such beliefs were untenable. This "as if" strategy was precisely what Nietzsche, a century after Hume and Kant, characterized as evidence of the nihilism in European culture: humans basing their lives on ideas that they knew, nonetheless, to be false.

The skepticism of Hume meant, in effect, an alienation of philosophy from real life and a marginalization of the role of the philosopher. That is, how could anyone who took Hume's thought seriously find a philosophically meaningful basis for his existence? Great classical thinkers like Plato and Aristotle, in contrast, claimed to discover through their reason higher truths that would serve to orient life and culture. Hume, however, called God, science, and reason into question and then serenely counseled humans to behave, in effect, as if his critique didn't matter.

In the end, what Hume had to offer as a practical guide for living was conformity to habit and custom—a conservative approach that precluded any serious critique of existing social and political arrangements. In his treatment of philosophy as an essentially academic activity without any responsibility for such critique, Hume was a precursor of the analytical philosophers and deconstructionists of the twentieth century who helped to shape the nihilistic ideological framework in which our present postmodern culture of technology without limits has emerged.

Kant's flawed attempt to rescue God, morality, science, and metaphysics from Hume's critique. Kant's philosophy was an attempt to rescue science from Hume's critique of the rationality of causation, morality from Hume's assertion that it was based on passion, and metaphysics and religion from the nihilism where Hume's skepticism had left them. In spite of these intentions, however, Kant's efforts involved further concessions to the attack on reason and had the eventual effect of deepening the crisis of European culture. Ultimately, his philosophical system relied heavily on faith and gave new importance to will. And it was will, rather than reason, which would come to play a decisive role in the elaboration of the reactionary modernist rationale for technology without limits that emerged early in the twentieth century.

Despite his famous demonstration in the *Critique of Pure Reason* (1781) that reason alone was unable to found a reliable metaphysics that would enable humans to know the ultimate truth of the world (i.e., the *noumena*, or "things in themselves"), and faced with his inability to prove either rationally or by experience the existence of a supreme being either inside or outside the phenomenal world, Kant offered, nevertheless, an alternative to Hume's skepticism. It was a "rational faith" in God that, he argued, would have the "regulative" merit of providing a deeper meaning for scientific investigations and, lacking any metaphysical certainties, a greater unity for philosophical concepts.[48] It would also, combined with faith in the existence of an immortal soul, provide a justification for the sacrifice one had to make in life of his or her individual desires in order to lead a moral existence in this world.

Kant's thought, in any case, was an expression of the post-Apollonian subjective philosophy that Descartes had originated in the seventeenth century with his "*cogito ergo sum*." In the first place, as part of his "Copernican revolution" in philosophy, Kant asserted that the mind of the individual subject (i.e., his *Verstand*) "constituted" the phenomenal world of nature by *bringing* such concepts as time, space, and causation *to* the sensuous manifold of its experience. Second, the subject's autonomous, rational will (*Vernuft*) was for Kant the basis of morality and a choice to have faith in God. That is, in contrast to the cause-and-effect necessity of the phenomenal world of nature, the moral actions of humans were based on their willing freely to obey a higher, rational law rather than yield to the dictates of their natural inclinations. This law was his famous "categorical imperative," according to which an action was moral only if it could be justly applied to humans in general and have the effect of treating them as ends, not means. And thirdly, the subject's own feelings, as we shall see, were the raw material upon which aesthetic judgments (*Urteilskraft*) were to be founded.

The problem with Kant's philosophical system, however, was that he separated these areas of cognitive truth: science, morality, and art. According to Jürgen Habermas, before Kant published his *Critique of Pure Reason*, religion had provided a basis for uniting these three areas of concern. Once Kant disassociated faith from knowledge, however, the unifying function of religion was lost. The tendency thereafter was for fragmentation in intellectual culture; and the problem for philosophers was how to bring back together what Kant had split asunder.[49] Given such a development in philosophy, it was not surprising that a theory of morally autonomous technology, severed from external values, emerged within a century.

INCREASING CONFIDENCE OF THE *PHILOSOPHES* IN SCIENCE AND SCIENTIFIC METHODS

The self-validating aspect of science. In the face of the unsettling cultural and ideological trends of the eighteenth century, more humans placed their faith in the tangible benefits and empirical methodology of the new science. On a popular level, Berkeley's idealist philosophy and Hume's skepticism, both of which undermined the epistemological pretensions of modern science, had no appreciable effect. On the contrary, science's positive, self-validating aspect— that is, its successful technological applications for the mastery of nature— amplified the number of converts to its methods and claims to be capable of discovering truth.

As the new science of thinkers like Bacon, Descartes, Galileo, and Newton gathered momentum in the eighteenth century, it presented a formidable challenge to accepted truths, many of which were still nothing more than received ideas, about humans, nature, morality, and God. The new science not only provided a form of truth that could make a valid claim to universality, but mathematical rigor and precision, the requirement that theory had to be validated in practice to satisfy the new positive standard of power, and the idea that all truth had to be purged of emotional bias and dogmatic influence combined to create a powerful basis for cultural change and ideological transformation. It was a formula designed for action in the world rather than discovery of higher philosophical truth.

A number of the leading *philosophes* lent their voices to this pro-science chorus. Voltaire, for example, expressed a faith in the universal validity of science when he stated that, in contrast to religion, there were no sects in science.[50] In his lectures on progress at mid-century, Turgot expressed the belief that every scientific innovation would automatically result in human betterment and that in the future, there would be a universal, positive culture based on science. And four decades later, Condorcet echoed the same one-sided and confident view in his *Sketch of an Historical Tableau of the Progress of the Human Mind.*

The perceived "lag" of the humanities behind scientific advancement. There were a number of reasons for the growing acceptance of science as the way to truth and fulfillment. A belief in the primacy of science was reinforced by the idea that the political knowledge and moral knowledge, which had the traditional role of controlling it, were unable to keep up with the more rapid pace and expanding scope of scientific advances. According to some *philosophes* who promoted the new theory of progress, like the Abbé St. Pierre, there was a cultural lag between the latest scientific knowledge and the more traditional artistic, moral, and social disciplines. Their recognition of this dichotomy reflected the traditional Christian division of truth between the Book of Nature and the Book of Scripture and Pascal's distinction between the analytical and simplifying mentality of the scientist and the more complex perspective of the philosopher. Some leading Enlightenment *philosophes*, however, openly expressed a greater confidence in the validity of scientific knowledge. Indeed, as far as religious truth was concerned, Laplace's mathematical demonstrations at the end of the century suggested that mathematical science could be used to demolish it.

In the Middle Ages, the prevailing view had been that scientific truth had to be consistent with religion. In the seventeenth century, Galileo and Bacon challenged that principle and fought to free science from theological control. Now, with a further decline of religious authority, impressive advances in the sciences provided by its new methods of investigation, and the appearance of a "lag" in the case of values, the burden of justification was essentially shifted to nonscientific varieties of thought, be they theological or humanistic. Henceforth, they had to demonstrate their validity—and often in terms of criteria that were external to their disciplines. This development tended to undermine further the Apollonian Greek idea that technology should be oriented and controlled by higher values.

Greater credibility of the precise mathematical formulations of science, despite the misgivings of some **philosophes.** Another point of attraction for science was its reliance on quantification. Some promoters of science even identified its mathematical character as the key to human betterment. Turgot, for example, stated that mathematics guaranteed the inevitability of scientific progress. Mathematics, he said, was the universal language of pure reason, cutting across all cultural barriers. It was precise and certain, whereas ordinary language was susceptible to distortions due to prejudice, superstition, passion, and emotion. With mathematics, Turgot argued, even moral knowledge could be removed from the disputes of the marketplace and translated into numbers and equations. Furthermore, he asserted, because of mathematics there would be no limit to progress. It would continue indefinitely, like an infinite progression in mathematics.[51]

Condorcet, who had studied mathematics and wrote the article in the *Encyclopedia* about it, believed that science, which he considered an essentially mathematical discipline, was the key to human betterment. In contrast to Turgot, however, he realized that ethics and sociology would never reach the level of certainty of physical science. In those disciplines, therefore, he called for a "social mathematics," based on the calculation of probabilities.[52]

In England, the utilitarian philosopher Jeremy Bentham fully adopted the approach that moral questions could be resolved mathematically. In 1789, he published his *Principles of Morals and Legislation*, in which he advocated a cost-benefit calculus for reducing normative questions to issues of fact involving quantified variables.[53] When technology became morally autonomous in the twentieth century, its pervasive efficiency calculus reflected this kind of quantitative reductionism, which is also the essence of the profit calculus of capitalism.

Eighteenth-century thinkers like Turgot, Condorcet, and Bentham, in any case, advocated extending the scope of quantifying thought to the realm of moral judgment. In doing so, they revealed a simplistic desire to absorb the humanities into the sciences and thereby eliminate entirely the methodological split between two forms of knowledge whose differences Pascal had so accurately described. The implication of such an approach was to suggest that no limits on science or technology could be valid unless they were obtained by the same epistemological methods science and technology employed. In effect, what Aristotle had called productive knowledge, concerned with means, was absorbing higher philosophical knowledge, devoted to ends. This was a step in the direction of the totalizing value system of modern technocracy, according to which everything is judged in terms of its effectiveness as means of mastery and control.

Diderot, however, was one of the *philosophes* who denounced the austere, mathematical language of Newton's *Principia* and recognized the potential of mathematics for mystifying a public not versed in science. It was a "veil," he said, that scientists were "pleased to draw between the people and nature."[54] Hume, too, did not share the faith in mathematics of thinkers like Turgot and Condorcet. He stated that it had no validity beyond that which imagination and will, influenced by habit, lent to its objects.[55]

For these and other reasons, a number of leading eighteenth-century *philosophes* recognized the danger in disconnecting science from philosophical control. Despite his support of science, Voltaire had misgivings. He wrote to the Abbé d'Olivet of the threat to life inherent in scientific neutrality and its "chilling objectivity."[56] Diderot, who sensed the increasing compartmentalization and one-dimensional focus of science wrote: ". . . [H]appy the geometer in which a consummate study of the abstract sciences has not weakened a taste for the arts."[57] Despite such disclaimers and declarations of misgivings, however, the forward march of a new, scientizing culture and civilization of technological means without end was acquiring an irreversible momentum.

THE EMERGENCE OF INDEPENDENT STANDARDS IN THE *TECHNE* OF ART

The trend toward subjective criteria. In the realm of fine art, the seventeenth century had passed on the legacy of neoclassicism to the eighteenth century. Its primary tenets were that art was governed by objective standards, that it had to aim at the moral improvement of the public, and that it had to be orderly and refined.[58] During the Enlightenment, however, the idea of objective standards was called into question. In France, Diderot called for a more irrational, subjective, and autonomous conception of the *techne* of art. Rousseau represented in his literary works a new subjective and sentimental standard; and Winckelmann, the German art historian, approached a work of art subjectively, through his homoerotic impulses.[59] In Great Britain, Burke, Hume, and Shaftesbury all were adamant in their opposition to purely rational criteria in art and promoted, respectively, passion, experience and opinion, and feeling.[60]

Kant's influential theory of autonomous aesthetic judgments. Kant refused to endorse either a purely rational or exclusively irrational explanation of beauty, although he clearly adopted a subjective approach. In his *Critique of Judgment* (1790), he stated that the determination of the beauty of a thing was based on a feeling of pleasure that was referred to the mind for judgment in a disinterested context of free interplay. That is, in contrast to the empirical judgments of science, which aimed at the mastery of nature, or the willful judgments of morality, which demanded the repression of natural impulses, Kant believed aesthetic judgments did not involve any desire to possess the object or imply the adoption of a particular moral attitude relative to its existence.[61]

Since aesthetic pleasure was disinterested, he argued, one who experienced it in the presence of a work of art had to assume it would be present in others in the same situation and that a judgment that the object was beautiful, therefore, would have universal validity. In taking such a position, Kant was essentially exhorting the aesthetic critic to make a judgment that would be universally valid.[62] Kant thus assigned the most important role in aesthetics to subjective reason, which had to determine, in effect, if a work of art had universal validity without the benefit of any objective concepts or guidelines.

The aesthetic of Kant, in any case, was not one of art for art's sake, although it did contain the seeds of such a conception. Primarily, it established a separate realm of cognition and truth for the *techne* of fine art, distinct from science and morality. As Gay put it, Kant separated art from the social and political thought of the period and gave it an independent path, like the science of the Enlightenment.[63] In this respect, he prepared the ground for the autonomous art of the second half of the nineteenth century, whose legitimizing principles were applied to morally autonomous technology by reactionary modernist and National Socialist thinkers in the twentieth century.

Formulation of a Modern Theory of Progress and Technological Development

If the roots of the modern theory of progress can be found in the linear and teleological Judeo-Christian concept of history enunciated by Augustine and the medieval idea of human improvement as recovery from the Fall, it was in the eighteenth century that this idea gained acceptance among members of the intelligentsia and began to seize upon the popular imagination as well.

France was the nation that provided the most fertile cultural terrain for the development of this new theory. It was the scene of the ideological showdown over cultural history between the "ancients" and the "moderns" at the end of the seventeenth century. And in the eighteenth century, it was the breeding ground for the ideas of progressivist thinkers like Turgot and Condorcet and the locus of the most important historical event of the century—the French Revolution. France was also the point of origin of the Napoleonic Wars, which changed the course of history for the rest of Europe.

FONTENELLE'S BRIEF FOR THE "MODERNS" OVER THE "ANCIENTS"

A decisive battle between the promoters of the present and the admirers of the past, the famous literary quarrel between the ancients and the moderns, took place in France at the end of the seventeenth century, when the Enlightenment was beginning. The moderns, whose point of view ultimately carried the day and set the cultural tone for the Enlightenment, were a literary clique who argued that the achievements of contemporary civilization were superior to those of classical antiquity. They pointed to the accomplishments of individuals like Newton, Galileo, and Shakespeare, as well as the brilliance of the reign of Louis XIV, which began in France in 1643. The ancients, whose adherents included Racine, Molière, and La Fontaine, maintained that the classical civilization of the Greeks and Romans had not been equaled in the present.

In his famous essay, *A Digression on the Ancients and the Moderns* (1681), Bernard de Fontenelle presented his completely secularized theory of the progress of human civilization. He based his argument on the development of the natural sciences, an area where the accumulation of knowledge over time led to advances and where the gains of the present were apparent and undisputable. A key reason for progress, Fontenelle maintained, was that the thinkers of the present benefited from the mistakes of the past and were able to draw on a greater collective human experience. Progress even occurred in the fine arts, he added, because each succeeding generation set higher standards than its predecessor. As long as governments favored the arts (which included both the fine and mechanical arts) and sciences, and wars and religious prejudice did not arrest the advance of civilization, he said, progress would continue.[64] The human race, Fontenelle stated confidently, would never reach "old age" or "degenerate."[65]

Like Pascal, however, Fontenelle stopped short of including the idea of human perfectibility. He did not believe morality was a progressive discipline, so he promoted the moral ideas of classical thinkers like the Stoics and the Epicureans.[66] In England, Locke's assertion that the human mind was a blank slate formed by experience suggested a possibility for human perfectibility, but he formulated no historical theory of inevitable moral improvement over time. This was a task that was left to the French *philosophes* of the next century.

VOLTAIRE'S REJECTION OF LEIBNITZ'S THEODICY

In *Theodicy* (1710), Leibnitz, a German, reasserted the traditional Christian view that God directly controlled every historical occurrence and expressed the optimistic belief that, as a consequence, "everything would turn out for the best in the best of all possible worlds" despite the widespread evidence of evil in the world. This was a far cry, of course, from the idea of gradual, continual, self-directed betterment of humanity.

In France, Charles de Montesquieu was one of the first *philosophes* to challenge this providential view of history in his *Spirit of the Laws* (1750). He argued that there were situational and cultural factors active in each society, and that when the influence of these factors was properly understood, peoples could determine their own destinies.[67] This was a modern view, similar to Bacon's approach in the physical sciences and Machiavelli's in the social sciences. That is, they all believed that once the laws that governed events were discovered, humans would be able to use such knowledge to influence the outcome.

Voltaire focused the historical analysis in his influential *Essay on Manners* (1756) on the progress of the human mind—that is, on what is often called the "history of consciousness." For him, history was a gradual process by which reason slowly overcame the obstacles of habit, custom, and prejudice. In each historical period, Voltaire sought to discover what he termed the culturally motivating "spirit of the times." He believed an understanding of history would assist the human mind (i.e., reason) to attain the goal of self-knowledge and self-consciousness, and the purpose of his *Essay* was to enable humans to advance toward this end.[68]

Voltaire disagreed with Leibnitz's providential view of history and argued in his *Essay* that progress could occur by secular means.[69] In his famous novel *Candide* (1759), he stated that God did not intervene in human history and that chance was often a determinant of historical events. He concluded, however, that it was up to humans to work to improve the conditions of their existence.[70]

THE DEFINITIVE VERSION OF PROGRESS, ELABORATED BY TURGOT AND CONDORCET

At exactly mid-century, the Abbé Turgot, Baron de l'Aulne, twenty-three years old, delivered two influential lectures in which he systematically laid out the new ideology of progress. The next year, Diderot published the first volume of his famous *Encyclopedia*, a project intended to facilitate progress by disseminating the new knowledge and ideas of the Enlightenment to its readers. When the French Revolution started in 1789, it led the *philosophes* to believe, according to Arendt, that progress was possible not only in knowledge but also in human affairs in a more general sense.[71] So strong was this conviction in Condorcet's case that he wrote his famous *Sketch of an Historical Tableau of the Progress of the Human Mind* (1793) despite the fact that the revolution was beginning to take an extreme direction and his own situation had become precarious.

Although Voltaire argued that the progressive tendency in history did not affirm itself until the eighteenth century—that is, the Age of Reason—Turgot and his biographer, Condorcet, believed that there had always been a progressive movement of history.[72] In Turgot's view, however, progress had been blind for most of the past, driven by the factors of passion (e.g., wars) and necessity. It had only become conscious in his own time, he said, when humans began to rely on reason.[73]

A guarantee of ultimate benefit, legitimizing all technological change. Turgot stated that there were two opposing tendencies always active in human behavior and history. One was a disposition to innovate and the other was a propensity to stagnate. It was the innovative tendency, he argued, that was responsible for human betterment. Regardless of any apparent risks or dangers, therefore, giving free rein to the innovation principle would yield the best result in the long run. He had read Leibnitz and believed, in effect, that providence would guarantee the outcome.[74] His motto was *laissez-innover*, meaning there should be no controls on innovation in science or elsewhere.[75]

Turgot's disciple and biographer, Condorcet, adopted a similar point of view, although he secularized the guarantee of a favorable result by replacing providence with history.[76] Every discovery in science, he reasoned, would turn out to be a benefit for the entire human race.[77] The science to which both referred, it should be noted, was of the Baconian and Cartesian variety, which yielded power knowledge capable of technological application.

This theodicy of *laissez-innover* meant that as far as technology was concerned, the theory of progress would become the new, secularized version of what Noble has appropriately called "the religion of technology," replacing its immediate but more sacral forerunner, the Baconian version of the millennium. Frank Manuel, in any case, labeled Turgot the "initiator" of the "rationalist prophetic tradition," which Condorcet and others more fully secularized.[78]

The assertion of inevitable benefit was essentially an appeal to faith, since Turgot relied on God's problematical existence and since neither he nor Condorcet, who excluded the God factor, could actually foresee the future. Drawing, most likely, on his perception of positive effects of a number of scientific discoveries and technological inventions during his own lifetime, Condorcet decided to assume that all innovation would lead to a similar beneficial result.

Given the irresistible appeal of the idea of inevitable benefit, the principle of *laissez-innover* is one of the main ideological reasons why technology development has essentially proceeded without concern for limits. In the contemporary world, scientists and technologists frequently make this self-serving assertion of inevitable societal benefit to counter any efforts to control their research, even if they are not familiar with Turgot and Condorcet nor the ideological climate that prevailed when they made their faith-based claim in the eighteenth century.

In spite of his assurances of a guaranteed positive result, however, Turgot admitted in 1750 that the fast and unremitting pace of change made it difficult for humans to understand correctly the implications of innovations in the here and now of the present. In a revealing and prescient statement, he admitted that "Before we have learned that things are in a given situation, they have already been altered several times. Thus, we always become aware of events when it is too late, and politics has to foresee the present, so to speak."[79]

Three goals of progress for Turgot. Turgot declared there were four goals that the human race would realize as it pursued its historical mission of progress. The first was to acquire and preserve an ever-increasing body of knowledge about itself and the physical world. This was an obvious reference to science, where knowledge was, indeed, cumulative. Fontenelle, who had based his theory of progress on advances in science, had likewise argued that the accumulation of knowledge inevitably led to human betterment.

The second goal was in the area of the fine arts, although Turgot made an important exception to the idea of continual improvement. Here, he stated, the goal could only be to reach a certain level of excellence and remain there. Artistic accomplishment, he explained, did not depend on the accumulation of knowledge, but rather on the imagination. Once perfection had been attained, he added, the nature and sensitivity of our organs prevented further advancement. Furthermore, he shared the neoclassical view, echoed by the ancients in their quarrel with the moderns in France, that the arts had reached a zenith in the Augustan age of Rome. He did not rule out the possibility, however, that a like level of accomplishment might be reached again.[80]

A third goal of progress, according to Turgot's analysis, was for humans to become civilized and moral beings. This idea constituted a break with the Christian idea of an irreparable flaw in human nature due to original sin and provided the basis for educational reform and progressive legislation in modern democratic societies. Condorcet gave this optimism its most forceful expression when he later declared that the potential for human improvement was without limit.[81]

Turgot's crucial fourth goal: technological advancement. According to Turgot, a fourth goal of progress was for humans to exert ever-increasing control over nature by means of technology. He asserted that technology was the area in which the boldest growth in human genius had occurred. Since technology served elementary human needs, was used by a vast number of people, and was not dependent on language, he reasoned, it was the least susceptible to loss and could survive even barbarian invasions.[82]

Turgot identified technological advancement as the most obvious and least questionable manifestation of progress. The chance of invention was greater than in other cultural domains, he said, because technology could be found everywhere on the planet.[83] Furthermore, his theodicy of *laissez-innover* implied that every technology ultimately would be used to serve positive goals and dispensed a priori with the possibility that some technologies might be more harmful than beneficial if their broader, longer-term impact were taken into account. In this respect, his conception of technology was already one of technology without limits.

In the future, Turgot predicted, faithful to the intent of Bacon's power knowledge, that technology would owe more to science than the contrary, although he stated that the regeneration of science in the recent past had been the product of technological advances of the Middle Ages. Such innovations included, he said, the magnifying glass, maritime instruments, and, most important, printing, which had served to disseminate knowledge broadly, make the discoveries of the Greeks available, and stimulate potential geniuses by making them aware of the achievements of their predecessors.[84]

A major goal of Diderot's *Encyclopedia, or Reasoned Dictionary on the Sciences, Arts, and Crafts* (1751-72) was, in the spirit of the new Baconian science, to make the mechanical arts a major component of the culture of the Enlightenment.[85] This goal was also consistent with the idea that technological development would be a primary factor in the progress and improvement of the human race. In 1777, Johannes Beckmann, a German, published his *Guide to Technology*, in which he confirmed its growing importance. He was one of the originators of the modern use of the term "technology" to situate the mechanical arts in a new unitary and autonomous branch of learning.[86]

This focus on technology, of course, was not surprising, given the fact that the industrial revolution had started and that it owed more at the time to technological innovation in its own right than to technological applications of Baconian science. These exclusively technological innovations included the steam engine and its spin-offs—steam-powered looms in the textile mills, steam-driven pumps in the mines, and steam-powered locomotives for rail transportation—as well as the new organizational techniques applied to the work process in the factory.

The Eurocentric bias in the theory of progress. Turgot believed that progress was inevitable as the scientific knowledge of Europe spread to all parts of the globe. The eventual result, he stated, would be a uniform and positive worldwide culture based on science. In the past, barbarian invasions had overwhelmed societies that had attained a higher cultural level; but in the present, Turgot asserted, scientific knowledge was so widely diffused that no further setbacks were possible. Furthermore, he said, in the backward areas that still existed, Europe had a mission to civilize the rest of the world. Its task was to incorporate all regions and all peoples into the scientific civilization of which it was already the avant garde.[87]

This Eurocentrism of Turgot was shared by Condorcet, who wrote in his *Sketch* that the Europeans residing in the colonies of the New World would "either civilize or peacefully remove [sic!] the savage nations who still inhabit vast tracts" of land. European colonists in Africa and Asia, he added reassuringly, would cease to be "tyrants or corruptors" and would become propagators of the "principles and the practice of liberty, knowledge, and reason that they have brought from Europe."[88] In the future, he blissfully declared, all nations would "attain that state of civilization which the most enlightened, the freest and the least burdened by prejudices, such as the French and the Anglo-Americans, have already attained."[89]

The contemporary Mexican writer, Carlos Fuentes, has taken issue with the "religion of futurity" he saw embodied in the modern theory of progress, in particular because of its implications for non-Western peoples. Once the linear, teleological history of Christian thinkers like Augustine was secularized, he said, and once God was replaced by the laws of history to which Condorcet and his nineteenth-century successors referred, all historical occurrences, as long as they contributed to the spread of European values and practices, were seen in Manichaean (i.e., exclusively positive) terms.[90] Progress thus became an ideological weapon serving, along with the West's superior technology and claim to be worshipping the one true God, the goal of dominating and exploiting the peoples of other parts of the world.

Philosophes like Voltaire, Locke, and Condorcet, according to Fuentes, took human nature as they defined it in eighteenth-century Europe to be universal. This Eurocentric distortion prevented Westerners from seeing members of tribal societies in Africa, Brazil, India, or New Guinea as essentially anything but "other." Thus othered, he added, they were reduced to the status of objects, suitable for political subjugation, cultural transformation, and even physical eradication when their presence impeded the march of progress. Fuentes therefore concluded that the future-oriented time of the West was "violent time" for non-European peoples.[91]

PHILOSOPHICAL DISSENT CONCERNING PROGRESS

For the bourgeoisie, the social class that had assumed the leading economic and political role in the most advanced European nations by the end of the eighteenth century, the theory of progress served as an ideology that legitimized its power and the changes it was effecting in the natural and social world by directing the process of industrialization. And for scientists and technologists, the ideology of progress assigned them a privileged role in society. The principle of *laissez-innover* ruled out a priori any efforts to limit or control their research and provided a rationale for increasing governmental support of their activities.

Despite the compelling elements of the theory of progress, however, some of the leading *philosophes* cast doubt upon its validity. Montesquieu wrote in his *Persian Letters* (1721) that nations went from barbarism to civilization and eventually back to barbarism. Hume expressed the view that advances in "industry, arts, and trade" also increased the power of the government over its subjects. D'Alembert stated that progress did not produce happiness, with greater enlightenment often resulting in disillusionment. Voltaire wrote at one point, "Everything has its limits ... genius has but one century; after that, everything must deteriorate." Kant saw progress in morality as questionable; and Wieland, Germany poet and *belle-lettrist*, pointed out that the time of greatest cultural refinement was also when moral corruption was rife.[92]

Although Turgot expressed the belief that humans eventually would be totally reasonable,[93] and Condorcet wrote that in the future "men ... will recognize no masters but their reason,"[94] some *philosophes* expressed reservations about the capacity of most human beings for enlightenment. Hume said that the vast majority of humanity was governed by authority and superstition. Diderot saw the ordinary people as too stupid to be enlightened. Kant referred to the *Volk* as "consisting of idiots." And Voltaire referred to humans as "two-footed animals."[95]

Rousseau's Radical Critique of Progress and Technology

HISTORY'S TRUE STORY OF DECLINE IN HAPPINESS AND MORALITY

A number of leading *philosophes*, as we have seen, expressed their reservations about progress. The most thorough and influential critique of the theory, however, was made by Rousseau in his *Discourses*. In his *Discourse on the Sciences and the Arts* (1751), Rousseau argued that advances in the arts and sciences did not lead to greater happiness or improvement of morality. His basic thesis was that civilization was repressive and led to unhappiness; and he argued that peoples like the early Persians, the Germanic tribes, the Romans "in the time of their poverty and innocence," and the tribes of America were morally superior to his more civilized, eighteenth-century European contemporaries.[96]

Rousseau therefore asserted that the solution to humanity's misfortunes was not in the progress of civilization, but rather in a return to nature and to the virtue and simplicity of the past. He included this famous prayer in this *Discourse*: "All powerful God . . . deliver us from the knowledge and fatal arts of our fathers, and give us back ignorance, innocence, and poverty, the only possessions which can make us happy and are precious in Thy sight."[97]

In his *Discourse on the Origin of Inequality* (1755), Rousseau escalated his assault on the idea of progress. He declared that humans were happiest at the beginning of history, when they had lived a solitary, primitive life in the state of nature, guided only by two sentiments, preservation of the self and pity for others.[98] Furthermore, the innovating tendency that Turgot had said was responsible for progress was, in Rousseau's view, "the source of all misfortune" and had made the human "the tyrant over himself and over nature."[99]

In rejecting Turgot's and Condorcet's guarantee of benefit from all innovation, Rousseau pointedly excluded himself from the ranks of those who promoted the secular religion of progress, with its emphasis on science and technology. For him, the history of civilization was rather a story of "so many crimes, wars, murders, so many miseries and horrors . . . ," initiated not by the Fall as related in Genesis, but by technological innovations and the invention of the institution of private property.[100] With his dissent from progress, which amounted to a secularized version of Christian historical teleology, Rousseau prepared the ground for nineteenth-century philosophers of will like Schopenhauer and Nietzsche in Germany and Realist writers like Flaubert and Baudelaire in France.

In *The Social Contract* and *Emile*, both published in 1762, Rousseau proposed two possible alternatives to the unhappy and corrupt society in which he was living. *The Social Contract* outlined the first, a political restructuring of society combined with a lifestyle close to nature. *Emile* contained the second, guidelines for educating a child to live a solitary and self-sufficient life on the margins of a corrupt society that had not been changed. Significantly, neither of these alternatives assigned technology a positive role in the improvement of the human condition.

TECHNOLOGY AS A PRIMARY BEARER OF HUMAN MISFORTUNE

If technology had been for Christian theologians a means for humans to recover from the Fall and was for Turgot the most undeniable example and a primary generator of progress, Rousseau differed in linking it to human misfortune and moral decline. He thus originated a critique of technology at the very point where other influential *philosophes* made it a bedrock of modernity. In his *Discourse on the Sciences and the Arts* (1751), Rousseau included technology in the general category of the arts, which he said had corrupted humans and made them unhappy. For example, he condemned printing, which Turgot had singled out for special praise, as the "art" which made it possible to "eternalize the extravagances of human beings" like Hobbes and Spinoza, philosophers he pointedly mentioned as purveyors of dangerous misconceptions.[101]

In his description of humans in their early, "natural" condition in the *Discourse on the Origin of Inequality* (1755), Rousseau expanded on the negative consequences of technological development. He pictured these early humans as happy and robust creatures who led a solitary existence close to an abundant nature that satisfied all their needs. The only "instrument" they knew at that time, he said, was their body. Later, he added, when stimulated by circumstances, they began efforts to improve their original condition and gained a measure of power over nature.[102]

The earliest human condition, therefore, was not one of perfect knowledge and power over nature, as Bacon and the Puritan millenarians had interpreted Genesis to signify, but was rather life in a state of nature characterized by ignorance and reliance on feelings. In taking this position, Rousseau discretely acknowledged that he was contradicting Christian doctrine, which, he said, "orders us to believe that God himself removed humans from the state of nature immediately after the Creation"[103] and teaches us "that the first human received immediately from God knowledge and teachings."[104] According to Rousseau, however, the early humans actually were creatures who slept a lot and thought very little.[105]

He cautiously justified this description of the events of early history as based on "hypothetical and conditional reasoning," drawn exclusively from "the nature of humans and the creatures around them" and made in conformity with the predetermined topic of his *Discourse*, written for an essay contest sponsored by the Academy of Dijon.[106] The fact that he did not win a prize for it, however, in contrast to his earlier *Discourse on the Sciences and the Arts*, was evidence that he had not succeeded in avoiding controversy over a delicate topic.

According to Rousseau, whose preference was for the individual rather than society, there was a difference between good and bad technologies. The former were those that could be used by a solitary individual and had been developed while humans lived in the original state of nature. The latter were those technologies that appeared later and required the conjunction of other individuals.[107] To the first category belonged such inventions as fishing tackle, the bow and arrow, animal traps, fire-making, and cooking, all useful in the solitary mode of existence that natural humans originally enjoyed. Innovations like these not only served to establish superiority over animals, said Rousseau, but they gave humans a feeling of pride in their species.[108]

Bad technologies, according to Rousseau, included metallurgy and agriculture, whose invention, he said, "civilized" humans. "Iron and grain," he declared, were the catalysts that touched off a sad chain of events which, a secular equivalent of the Fall, brought down upon humanity a history of unhappiness and misfortune.[109] Metallurgy was invented, he explained, when humans found ways to imitate volcanoes that they had seen "vomiting out fused metallic substances." And as soon as a segment of the population was employed in forging and melting metals, he added, the invention of agriculture followed. Since the metalworkers were no longer engaged in providing for the common subsistence, agricultural technologies became necessary to produce enough food for them and the remainder of the population.[110]

The plowing and cultivation of fields, Rousseau continued, soon necessitated dividing land into private holdings, as well as the creation of a labor force without land of its own. And with the invention of the social institution of private property, he added, the strong accumulated property and became rich, while the weak ended up poor. These developments, according to Rousseau, magnified the natural inequalities of humans, rendered their differences more permanent, and increased their importance for the destiny of each individual. Class warfare began, with "usurpations" on the part of the rich and increasing robbery by the poor. The rich then tricked the poor, said Rousseau, into agreeing to the creation of a government to protect life and property, and inequality thus became legalized. This was the point at which, in Rousseau's secularized version of the Fall, the human condition became an unhappy one.[111]

He made it clear in his first *Discourse*, however, that recovery of the original happiness and virtue of humans could come only from a return to the innocence and simplicity of the past, not from progress in the arts and sciences, as Bacon and the Christian millenarians had believed. For Rousseau, science and technology were not the answer to human problems and aspirations but rather a source of human misfortune. He never subscribed, therefore, to what Noble calls the "religion of technology" in either its theological or secular form.

It is significant that Rousseau, who was not an atheist, broke nevertheless with the mainstream Christian views of both history and technology, secularized as they were by the eighteenth century. In making this break, he had some affinity with Jacques Ellul, a twentieth-century Augustinian Christian who rejected the idea that technological advance was a primary factor in the realization of human destiny and who made an extensive critical analysis of modern technology.

Confirmation, in any case, of Turgot's optimistic view of the effects of technological development and Rousseau's negative assessment could be found in different aspects of the industrial revolution that began in the eighteenth century. The new technological hardware, such as the steam engine, the mine pump, and the weaving machine, and the new organizational methods, like the routinization of repetitive and mindless tasks in the factory, made it possible to produce more and cheaper goods, not only for the growing middle class, but, in the more distant future, for the workers themselves. These innovations also led,

however, to pollution, environmental destruction, brutalization and dehumanization of the worker, and exaggerated differences in wealth between rich and poor
as industrial capitalism took hold and expanded its reach. The conclusion one
drew about the overall nature of progress, therefore, depended in most cases on
which consequences one chose to see and which consequences one chose to
ignore.

Notes to Chapter 4

1. Isaac Newton, *Newton's Mathematical Principles* (Berkeley: University of California Press, 1946), 547.

2. Ibid., 546.

3. Peter Gay, *The Enlightenment: An Interpretation*, vol. 2 (New York: Alfred A. Knopf, 1969), 170.

4. Roger Hahn, "Laplace and the Mechanistic Universe," in *God and Nature: Historical Essays on the Encounter Between Christianity and Science* (Berkeley: University of California Press, 1986).

5. Roger Hahn, "Laplace and the Vanishing Role of God," in *The Analytical Spirit: Essays in the History of Science in Honor of Henry Guerlac*, ed. Harry Woolf (Ithaca: Cornell University Press, 1981), 85.

6. Harry Prosch, *The Genesis of Twentieth Century Philosophy* (Garden City, N.Y.: Anchor Books, 1966), 292.

7. Gay, vol. 2, 162.

8. André Lagarde and Laurent Michard, ed., *XVIII Siècle: Les Grands Auteurs Français de Programme* (Paris: Bordas, 1953), 278-79.

9. Jean-Jacques Rousseau, *Dialogues, Rêveries D'Un Promeneur Solitaire (Extraits)* (Paris: Librarie Larousse, 1941), 22.

10. Jean-Jacques Rousseau, "La Nouvelle Héloïse," in Lagarde and Michard, *XVIII Siècle*, 284.

11. Rousseau, *Dialogues, Rêveries*, 38-40.

12. Paglia, 235-36.

13. Horkheimer and Adorno, 101.

14. Albert Camus, *L'Homme Revolté* (Paris: Gallimard, 1959), 57.

15. Albert Camus, *The Rebel* (New York: Vintage Books, 1963), 44.

16. Gay, vol. 2, 153-54.

17. Ibid., 155.

18. Ibid., vol. 1, 89.

19. Ibid., vol. 2, 170.

20. Antoine-Nicolas de Condorcet, "The Progress of the Human Mind," in *Heritage of Western Civilization*, 6th ed., vol. 2 (Englewood Cliffs, N. J.: Prentice Hall, 1991), 90, 94.

21. Ibid., 97.

22. Spink, 314.

23. Gay, vol. 1, 396.

24. Ibid., vol. 2, 527.

25. Ibid., vol. 1, 396.

26. Denis Diderot, Pensées Philosophiques; Lettre Sur des Aveugles; Supplément au Voyage de Bougainville (Paris: Garnier-Flammarion, 1972), 65.

27. Gay, vol. 1, 336-37.

28. Ernst Cassirer, *The Philosophy of the Enlightenment* (Boston" Beacon Press, 1955), 135.

29. Diderot, 147.

30. Notes taken by the author at a lecture by Prof. Wallace Matson, U.C. Berkeley, Spring, 1982.

31. Cassirer, 175-77.

32. Ibid., 170.

33. David Hume, *Dialogues Concerning Natural Religion* (New York: Hafner Press, 1948), 17-25, 33-34.

34. Cassirer, 179-80.

35. Hume, 64-67, 87-95.

36. Gay, vol. 2, 120.

37. Jean-Jacques Rousseau, *Discours sur les Sciences et les Arts. Discours sur l'Origine et les Fondements de l'Inégalité Parmi les Hommes* (Paris: Garnier-Flammarion, 1971), 168.

38. Gay, vol. 1, 402.

39. Russell, 670.

40. George H. Sabine, *A History of Political Theory* (New York: Holt, Rinehardt, and Winston, 1961), 600.

41. Prosch, 101.

42. Gerald J. Galgan, *The Logic of Modernity* (New York: New York University Press, 1982), 135.

43. Ibid.. 133.

44. Russell, 673-74.

45. Prosch, 110.

46. Russell, 673.

47. Stanley Rosen, *Nihilism: A Philosophical Essay* (New Haven and London: Yale University Press, 1969), 66.

48. Prosch, 160-61.

49. Jürgen Habermas, *The Philosophical Discourse of Modernity* (Cambridge: MIT Press, 1987), 18, 20.

50. Galgan, 147-48.

51. Frank Manuel, *The Prophets of Paris* (New York: Harper Torchbacks, 1962), 40, 43-45.

52. Antoine de Condorcet, *Esquisse d'un Tableau Historique des Progrès de l'Esprit Human* (Paris: Editions Sociales, 1966), 27, 67.

53. William Ebenstein, *Great Political Thinkers*, 3d ed. (New York: Holt, Rinehart and Winston, 1960), 500-01.

54. Gay, vol. 2, 158.

55. Rosen. 67-70.

56. Gay, vol. 2, 159.

57. Ibid., 158.

58. Ibid., 219

59. Ibid., 293-4.

60. Ibid., 303-10.

61. Paul Edwards, ed. in chief, *Encyclopedia of Philosophy*, vol. 4 (New York: MacMillan Company and The Free Press, 1967), 319.

62. Gay, vol.2, 317.

63. Ibid., 317-8.

64. Xavier Darcos and Bernard Tartayre, ed., *Le XVII Siècle en Lettres* (Paris: Hachette, 1987), 366.

65. A. Chassaig and Charles Senninges, ed., *Recueil de Textes Littéraires Françaises, XVII Siècle* (Paris: Hachette, 1975), 438.

66. Gay, vol. 1, 317-18.

67. Cassirer, 213-14.

68. Ibid., 216-19, 220.

69. Lagarde and Michard, ed., *XVIII Siècle*, 155-57.

70. Ibid., 160-61, 167-69.

71. Arendt, Part 2, 154.

72. Manuel, 34-35.

73. Ibid., 17.

74. Ibid., 47-48.

75. Ibid., 23-26.

76. Ibid., 64.

77. Monique and Francis Hinker, "*Introduction: La Vie de Condorcet*," in *Esquisse d'un Tableau Historique du Progrès de l'Esprit Humain*, 27.

78. Manuel, 16.

79. Ibid., 22.

80. Ibid., 17, 38-39.

81. See note 20, above.

82. Manuel, 37.

83. Ibid.

84. Ibid., 37-38.

85. Lecture by Jan Sebestik, of L'Institut d'Histoire des Sciences, Paris, at University of California, Berkeley: "The Beginning of Technological Thinking in the Late Eighteenth and Early Nineteenth Centuries," 1 Dec. 1981.

86. Ibid.

87. Ibid., 42.

88. Beatty and Johnson, ed., *Heritage of Western Civilization*, 6th ed, vol. 2, 91-2.

89. Ibid., 91.

90. Carlos Fuentes, "Writing in Time," *Explorations* (Jan. 1982): 62, 67.

91. Ibid., 63-66.

92. Gay, vol. 2, 101-07.

93. Manuel, 48-49.

94. Gay, vol. 2, 120.

95. Ibid., 518-9.

96. Rousseau, *Discours*, 43.

97. Ibid., 57.

98. Ibid., 197.

99. Ibid., 171-72.

100. Ibid., 205.

101. Ibid., 57.

102. Ibid., 164, 171-72.

103. Ibid., 159.

104. Ibid., 158.

105. Ibid., 170.

106. Ibid., 158-59.

107. Ibid., 213.

108. Ibid., 206.

109. Ibid., 213.

110. Ibid., 214-15.

111. Ibid., 215-20.

5
THE NINETEENTH CENTURY

Competing Philosophical Paths to Truth and Fulfillment in the Nineteenth Century: Faith, History, Reason, Will, Science, and Art

There were a number of significant efforts to revitalize philosophy and provide a way to give higher meaning to life and culture in the nineteenth century; but by its end, the advance of nihilism had not been checked. In effect, the epistemological implications of the revolution in science, Kant's separation of scientific, moral, and aesthetic truth, Hume's radical skepticism, and the failed attempts of philosophers to provide a satisfactory rational or irrational basis for belief in God proved to be a difficult, if not impossible, legacy to overcome. Each new attempt to find an answer to the cultural crisis in the West proved to be unsatisfactory, as Nietzsche's fin-de-siècle critique of historicism, science, art, philosophy, and religion demonstrated.

HISTORY AS THE SOURCE OF TRUTH AND THE VEHICLE FOR REALIZATION OF HUMAN DESTINY

Hegel and Marx and Engels were the last great European philosophers to adhere to the Judeo-Christian historical tradition, although Marxist philosophy was exclusively secular. They believed, like Augustine in the fifth century, Bacon and the British Puritans in the seventeenth, and Turgot and Condorcet in the eighteenth, that human destiny would be realized as the historical process unfolded. In other words, they accepted the idea that history was a linear and teleological process.

There was an important difference between these two nineteenth-century historicist conceptions, however. Whereas Hegel saw history as the story of a progressive development of ideas and consciousness, Marx and Engels, like the Christian millenarians of the late Middle Ages and Renaissance, saw a relation between technological advance and historical progress.

Hegel's history of consciousness. In his *Essay on Manners* (1756), Voltaire had described history as a process by which the human mind would attain self-knowledge and self-awareness. This idea had an obvious influence on Hegel, whose idealist philosophy in the nineteenth century was an account of how Reason (i.e., the Absolute Spirit) had attained freedom in a historical process of becoming conscious of itself.

According to Hegel, nature was a creation of Mind. True reality for him was consciousness, or self-thinking thought, which he called the Absolute Spirit. This was his philosophical equivalent of God, which he emptied, like Aristotle, of all anthropomorphic qualities except thought itself. Spirit's true essence, he said, was freedom, whose realization was the "absolute, final purpose of the world."[1] It could only be realized through self-recognition, however; and to achieve it, Spirit had to be embodied.

Spirit, therefore, initially "being-in-itself," objectified itself by creating the phenomenal world of nature. Humans were part of this world of nature, but they also participated in Spirit because they had reason. Spirit first emerged from nature, he said, when humans put Mind, by means of their labor, into nature. At this point, the history of consciousness began. Spirit then began a slow process, immanent in the movement of history and ultimately realized by the agency of human reason, of rising to self-awareness and freedom as "being-in-and-for-itself." It rose by historical stages, he added, to participate finally in the self-knowledge of the Absolute Spirit. This was the point at which his own idealist philosophy explained this process and showed it was the whole purpose of existence.

Before this terminal point in the history of consciousness was reached, however, Hegel said that different national cultures, some of which were non-European, had occupied the forefront of the historical process at different times.[2] Each produced the philosophers who advanced thinking toward the final goal of self-awareness. For Hegel, the conflicts among nation-states and cultures were an essential part of this historical dynamic. The nation-state, he said, was the only place where the idea of freedom was able to become an objective reality.[3]

In Hegel's system, all progressive social changes were ultimately the result of dialectic changes in thought. He explained how each philosophy carried within itself an opposing and negating tendency that eventually led to a synthesis at a higher level of understanding.[4] Within a national culture, its laws, religions, and morality, as well as the institutions that embodied them, advanced through a similar dialectic process of continual internal tension and adjustment.[5]

Like Augustine's and Leibnitz's providential explanation of history and Turgot's theory of progress, Hegel's philosophy of history included a theodicy. That is, although individuals acted without consciously seeking to serve the realization of the historical process, their actions, good or bad, ultimately had that effect. This was because the omniscient wisdom of the Absolute Spirit, which he called the "cunning of Reason," ensured that each individual's actions, regardless of the intention or objective, ultimately contributed to the attainment of the final goal.[6] This explanation had obvious similarities with Augustine's assertion that although humans had free will, God had foreknowledge of their actions, and history, therefore, was the unfolding of His plan.

In contrast to other Western thinkers, who were faithful to the Judeo-Christian belief that human destiny would eventually be historically realized, Hegel considered his own philosophy to be the final accomplishment of history. That is, it enabled Reason to become reconciled with itself by explaining history as the process by which it had risen in progressive stages to attain self-awareness. Thus Hegel stated in a famous dictum, "The owl of Minerva spreads its wings and flies only with the falling of dusk."[7] This meant that truth could only be gained from hindsight.

The task of the philosopher, according to Hegel, was only to be an interpreter of the times, not to offer advice or panaceas. His focus was on the past, not the present or the future. And since all historical outcomes were the product of Divine Reason, they were morally right. In contrast to Marx and Engels, who proposed action to change the world and saw the goal of history, communism, as yet to be realized, Hegel proposed acceptance of the "rational"—that is, what had historically already come to be.

In the context of Western thought, Hegel expressed a preference for the traditional values of German *Kultur* over the abstract principles of French *Zivilisation*. His experiential basis for this choice could undoubtedly be found in the real-life events of the Napoleonic Wars, in which the German states fought against the French invaders. Hegel rejected, in any case, abstract, yet-to-be-realized ideals of the Enlightenment such as equality and democracy, which he denounced as the basis of the aberrant individualism of the French revolutionary era and the source of the illusion that humans could remake society at will.[8] In contrast, Hegel defended those ideas, institutions, myths, and traditions that belonged to the category of *Kultur*, meaning they had developed organically over time and were therefore positive.

Furthermore, it was only a higher, synthetic form of reason (*Vernuft*) of the kind he employed in his philosophy to explain everything in terms of a totality, not the inferior reason of the understanding (*Verstand*) that was able to explain correctly the historical process and show that what had come to be was right.[9]

Verstand, which served only to break up organic wholes into unrelated and abstract components useful for manipulation and control, could not for Hegel provide true knowledge.[10] Science and technology, significantly, belonged to this category and could have only a secondary role.

Hegel's articulation of this *Zivilisation-Kultur* dichotomy had been anticipated in some respects by Rousseau's attack on reason and civilization and his stated preference for rural life, the organic community, and simple virtues. In choosing the positive values of *Kultur* over the abstract principles of *Zivilisation*, Hegel was a precursor of Nietzsche and the German reactionary modernist thinkers of the twentieth century. The latter, however, in contrast to Nietzsche, glorified technology and contrived to link it, despite its natural affinity to *Verstand*, with the traditional values of German *Kultur*.

Marx and Engels's materialist dialectic of history. Successors to the Enlightenment tradition of the French *progressistes*, Marx and Engels claimed, in effect, to have discovered the secular laws that governed history by analyzing the factual record of the past. In contrast to idealist philosophers like Fichte and Hegel, however, Marx and Engels said the underlying truth of human existence could only be found in production and productive relations—that is, in the material conditions of real life. Marx asserted, accordingly, that the "life processes," the techno-economic activities in which humans engaged to produce the material goods necessary to sustain their existence, were the essential species-defining characteristic of humans.

Furthermore, Marx and Engels also believed that the development of productive forces, not ideas, determined the progressive movement of history and gave it its ultimate meaning. Thus Marx wrote in his *German Ideology*, "If in England a machine is invented, which deprives countless workers of bread in India and China, and overturns the whole form of existence of these empires, this invention becomes a world-historical fact."[11]

As this production-driven historical process unfolded and humanity attained ever-more advanced levels of technology and production, it moved toward its point of realization, the victory of communism. At that point, Engels wrote, humans, for the first time in history, would become "the real, conscious lord of nature."[12] Such a declaration, of course, had an obvious affinity with the Judeo-Christian nature ethic of Genesis.

Like Hegel, Marx and Engels found there was a dialectic immanent in this process. It operated so that the techno-economic elements of the existing social system, the thesis, advanced to a level of development where they qualitatively no longer belonged to it, creating an antithesis. Contradictions developed, undermining the existing system from within and preparing the way for the transition to a more advanced level of social existence, the synthesis.

These progressive historical stages of social development included barbarism, tribal society, classical civilization, feudalism, and capitalism, each corresponding to a more advanced technological level of production and a more advanced level of civilization, except in each instance there was immorality and injustice because of the exploitation of the working class by a ruling class. Marx and Engels called upon the industrial proletariat in the capitalist phase, therefore, to carry out a revolution that would usher in the victory of communism. At that point, classless society would be established and human existence would be completely transformed for the better.

"Life is not determined by consciousness," Marx pointedly wrote in *The German Ideology*, in disagreement with Hegel's idealism, "but consciousness by life."[13] For Marx, therefore, unless one were familiar with his philosophy and accepted its explanation of reality, the consciousness so central to history according to Hegel was—and had always been—a false consciousness. The dominant bourgeoisie in the capitalist phase, for example, erroneously believed, according to Marx, that the existing forms of government, laws, morality, and social institutions were "eternal" precisely because it did not know that when the tools, machinery, methods, and organization of production underwent a qualitative change, everything else would have to change as well.[14]

Like the Puritan idea that technological advance would hasten a Second Coming of Christ and mark the onset of the millennium, Marx and Engels believed the arrival of advanced technology—in this case industrial production—would make possible the creation of a just and happy society on earth. For them, however, it would not be the millennium dear to some Christians and Jews but communism, a radically better and qualitatively different society in comparison to all that had preceded it.

As Engels explained in his *Socialism: Utopian and Scientific* (1891), in a communist society the human being would become "the real, conscious lord of nature ... make his own history ... [and] ascend from the Kingdom of Necessity to the Kingdom of Freedom."[15] This triumph of freedom over necessity was the secular and terrestrial equivalent of Augustine's Kingdom of the Saints in Heaven, where the immortal souls of the Just would enjoy a freedom where they could no longer sin. That is, at the end of history, when God destroyed the world, there was to be, according to Augustine, a "second resurrection" of the bodies of the Just, who would "enjoy one another's beauty without lust."[16] Desexualization of the body in heaven would be a triumph of Christian freedom, whose abuse had led at the very outset to Adam and Eve's loss of sexual innocence and expulsion from the Garden of Eden.

Engels thus cleverly borrowed from the language of Augustinian eschatology to describe, in secular but not totally unfamiliar terms, the radical transformation of the human condition on earth, which the victory of communism would bring. In the Marxist sense, the conquering of external nature made possible by technology would bring freedom. In Augustinian terms, however, it was the transformation of the human nature of the Just in heaven that would bring its realization.

WILL AND WILLING AS THE ESSENTIAL TRUTHS THAT EXPLAIN EXISTENCE

In the preceding century, both Rousseau and Kant had emphasized the importance of will, Kant for the purpose of morality and Rousseau for the purpose of communal legislation. There were a number of nineteenth-century thinkers who continued in this direction and focused on will as the key to truth and human fulfillment. Fichte and Nietzsche gave will a positive role, whereas Schopenhauer's philosophy treated it negatively and saw fulfillment in escaping its never-ending demands. In the twentieth century, German reactionary modernist philosophers went a step further by linking technology with the personified concept of will so that as a phenomenon it took on a reified status.

Fichte's philosophy of moral willing. According to Fichte's idealist philosophy, set out in his *Vocation of Man* (1800), nature was a creation of the Absolute Ego, which he described as a disembodied entity engaged in endless strivng to realize its vocation of free moral willing. It posited non-ego, or nature, as an obstacle to overcome. Nature, therefore, was only a means, a limit and field in which the ego could attain self realization.[17]

In Fichte's world, the finite egos of individual humans shared with the Absolute Ego a moral impulse to perform certain actions for the sake of performing them. The convergence of their actions toward an ideal goal contributed to the fulfillment of the Absolute Ego's striving.[18] Since they, too, needed determinate objects to overcome, nature provided them also with a necessary means for the realization of their actions. It was not a "foreign element," he emphasized, adding that in the case of the individual ego, nature was molded by that ego's laws of thought and had to be in harmony with them.[19]

Fichte followed Kant's lead and emphasized rational moral willing, which he said was the "vocation of man." He was essentially a romantic thinker, however, because he attached great value to yearning and striving, even if the goals one willed were never successfully attained. "Should it appear that during my whole earthly life," Fichte stated, "I have not advanced the good cause a single hair's breadth in this world, yet I dare not cease my efforts: after every unsuccessful attempt, I must still believe that the next will be successful."[20]

To make such a position tenable, however, he admitted that those who subscribed to his philosophy had to base it, ultimately, on faith. In the first place, he insisted, the finite ego of each individual had to have faith in a future transcendent life, since it would not always be apparent in this world that the consequences of a virtuous action would be beneficial for him. Second, it was also necessary to have faith that the external world, the phenomenal world of nature, which he said the Absolute Ego had "posited," actually existed.[21] And third, he argued that in order to believe in an Absolute Ego, faith in God was necessary.[22] Thus, Fichte concluded, "Our present life is a life in faith."[23] The problem, as Nietzsche subsequently pointed out, was that if one lost this faith, there would be nothing left except nihilism. As Santayana observed in the last century, the true object of Fichte's Absolute Will was willing itself.[24]

Fichte was, nevertheless, also firmly situated in the Judeo-Christian tradition of linear, teleological history. That is, he believed that a multiplicity of individual vocations of moral willing would converge toward the establishment of a universal world order. Recognizing in 1800 that this order had not yet been attained, however, he declared that the unification of the German people into a single, national state would serve in the interim to preserve for them a system of rights. And when humanity finally attained full moral development, he added, the state as an institution would wither away.[25]

Schopenhauer's pessimistic theory of the will. The first important nineteenth-century thinker to openly reject the historicist tradition was Schopenhauer. He published his doctoral thesis in 1833, *On the Will in Nature* in 1836, and *The World as Will and Idea* in 1844. Arguing that there was no plan or direction in history, he stated that "the human world was the kingdom of chance and error."[26] To underscore the importance of his position, he deliberately scheduled his lectures at the University of Berlin to coincide with those of Hegel. When his classrooms remained empty, however, he resigned in disgust.

Schopenhauer wrote in his famous philosophical work *The World as Will and Idea* (1844), that "[c]onsciousness constitutes the whole world as idea . . ."[27] Although he agreed with Hegel that the external world—that which he said was "of representations"—was the creation of the mind, he was not a philosophical idealist. That is, he agreed with Kant that there was an ultimate reality of the thing in itself.

For Schopenhauer, however, this reality was not outer, but inner. It was a striving, blind, and animal impulse, an irrational will to live, without affinity to Fichte's or Kant's rational free will. Individual wills were merely manifestations of it.[28] It was even anterior to instinctual desire, Schopenhauer said, because the body was the objectification of will as revealed by such drives as hunger and sexual lust.[29]

Because of this restless will, Schopenhauer saw all natural, and human, existence in a pessimistic, conflictual vein. "The will," he wrote, "dispenses with a final goal. It always strives . . ."[30] The result, he said, was "that the multitude of natural forces . . . everywhere strive with each other . . . a constant internecine war is waged . . ."[31]

Furthermore, because of the will's continual lack of any lasting satisfaction, human beings required *panem et circenses* (bread and circuses) and "all human life was tossed backwards and forwards between pain and ennui."[32] In other words, boredom was, for Schopenhauer, an existential dilemma rooted in human nature itself. The question, therefore, was how to escape the despair and frustration that it brought to all existence.

Just as he saw love as only transitory and providing only temporary satisfactions for the restless will, Schopenhauer did not consider suicide, a more radical "solution" that involved deliberate termination of one's existence, to be a viable alternative. Although it would eliminate a particular, individual will, he argued, the universal will to live, which was the cause of all the misery in the world, would continue to operate and exist intact.[33]

For Schopenhauer, there were no collective answers to the problem of the will. And the only viable way for the individual to be free from the tyranny of the will, outside of asceticism, was artistic contemplation. A work of artistic genius, he said, "reproduced the eternal [i.e., Platonic or Kantian] Ideas grasped through pure contemplation" and removed them from the "stream of the world's course" so they could be viewed independently of all relations.[34] With art, Schopenhauer argued, the knower functioned not as an individual consciousness, but as a "pure will-less subject."[35] Art thus had an exalted role for him, and he treated it like a supreme value, as did some of the greatest artists and writers of his century.

Dostoevsky's "Sniveling Hero": Nihilistic promotion of the unfettered will. There were some thinkers in the nineteenth century who portrayed humans as having an irresistible need for activity that they would seek to satisfy at any cost, without any reference to a historical context and without any concern for actual concrete results. In *Faust* (1801, Pt. 1), for example, Goethe described the protagonist as ready even to make a pact with the devil in exchange for secret powers that would serve his boundless and ruthless appetite for life and experience. Thus, when Faust declared, "Restless activity proves the man," Mephistopheles replied, "For you, no bound, no term is set."[36]

Similarly, in *Notes from the Underground* (1864), Dostoevsky's Sniveling Hero expressed his opposition to the scientizing positivist values of European culture at the time. Such values were symbolized for him by the Crystal Palace, the iron and glass showpiece of the World Exposition at Hyde Park in 1851, where the latest scientific and technological innovations were proudly exhibited.

Because of the restlessness of the will, Dostoevsky's Sniveling Hero objected, humans would not want to live in the Crystal Palace once they had completed the task of building it.[37] He proclaimed that a human being was a creature not only of reason, but also of will. Echoing such earlier thinkers as Bruno, Della Mirandola, Pascal, and Schopenhauer, he said a human had a "frivolous" and "incongruous" nature that led him, "like a chess player," to love "the process of the game, not the end of it."[38] He was "preeminently a creative animal," the Sniveling Hero added, "predestined to strive consciously for an object and to engage in engineering—that is, incessantly and eternally to make new roads, wherever they may lead."[39]

To prove that a human being was not just something rational and predictable like a "piano key," the Sniveling Hero said, he was willing, "at the cost of his skin ... to launch a curse upon the world ... to introduce ... his fatal, fantastic element."[40] This declaration of indifference to the consequences of one's actions, no matter how drastic, as well as the affirmation that only process and striving, not results, really mattered, had, of course, nihilistic implications.

Ironically, despite Dostoevsky's hero's rejection in this text of a culture based on the positivist values of an objectifying and standardizing science, his words were cited in 1975, more than a century later, by Samuel Florman, a publicist for the engineering profession, in a not-unfamiliar effort to create a technological imperative out of human restlessness and a perceived need to engage in constant activity. Quoting approvingly the Sniveling Hero's description of human beings as inherently restless, willful, and indifferent to consequences, he responded to contemporary "humanist" critics that placing any limits on technological innovation and development would be contrary to human nature and therefore an unacceptable infringement of human freedom.[41]

ART AS THE ANSWER TO THE MEANINGLESSNESS OF HUMAN EXISTENCE

After the disappointments in France and elsewhere in Europe of the Revolution of 1848, a number of the leading literary figures and artists of the time subscribed to Schopenhauer's rejection of history and the idea that art provided a way to escape the boredom and meaninglessness of existence. By means of their works and pronouncements, they promoted the idea that art was the answer—at least for those who had access to their works and a sufficient amount of leisure and level of cultural awareness to be able to reflect on the shortcomings of the times.

In France, for example, Gustave Flaubert, one of the greatest literary figures of the century, expressed feelings of boredom and contempt for society, echoed Schopenhauer's cynicism about love, and saw art as the only valid alternative. He related how, as a youth, he had had a premonition of the unhappiness of existence, like a "nauseous kitchen odor which made him sense he would vomit if he ate the food."[42] And at age twelve, he had referred to life as a "poor joke."[43] As an adult, however, he wrote to his mistress, Louise Colet: "Art is the only good and true thing in life. How can you compare a human life with it?"[44]

His famous *Sentimental Education* (written 1843-5 and published posthumously in 1910) was intended to be the story of a generation that had been betrayed by history—or more specifically, by the failure of the Revolution of 1848 to bring any positive change in people's existence. The book's main theme was failure—not only of history, but also in the life of his protagonist, Frederic, whose great love affair with a married woman was never even consummated.

Another famous Flaubert work, *Madame Bovary* (1857), was the story of the depressing existence of an unhappily married woman. Fed up with the tedium and emptiness of her life as the wife of a dull, provincial doctor, Emma Bovary embarked on several adulterous love affairs. They all ended unhappily, corrupting everything around her and eventually bringing about her own destruction. Despite, however, the depressing content of both of these novels and Flaubert's own expressed contempt for his characters in *Bovary* (whom he described as "profoundly repulsive"), he succeeded admirably in his primary intention, which was to write a work whose real meaning was in the beauty of its form—that is, in its literary style, or art.

In music, the German composer Richard Wagner, who in 1854 had met Schopenhauer, was a firm believer in this new religion of art. And in poetry, Baudelaire declared his boredom with existence and his contempt for conventional society, but a love for art. In the first poem of his famous collection *The Flowers of Evil* (1857), he singled out ennui as "the most ugly, mean, and vile"—and ubiquitous—of human vices. It could "willingly turn the earth into debris," he said, and "swallow the world in a yawn."[45]

Like *Madame Bovary*, Baudelaire's *Flowers of Evil* was a collection of beautiful writings about unsavory subject matter, including vice, crime, drugs, death, sin, hate, etc. The beauty of the form was what counted for him, according to the new Realist aesthetic he professed. And in a famous prose poem, "The Stranger," he declared his contempt for family, said he did not know the meaning of the word "friends," confessed to being ignorant of where his country was located, and said he hated gold "like you hate God." In contrast, he praised beauty, an "immortal goddess" whom he said he would "love . . . willingly."[46]

SCIENCE AS THE SOURCE OF TRUTH AND THE MEANS TO HUMAN FULFILLMENT

The incompatibility of Darwin's theory of evolution and Genesis. Darwin's theory of evolution, which was based on close and detailed observation of nature during a five-year voyage on the ship *Beagle,* provided support for those who claimed that science, not religion, was a more credible source of truth. Furthermore, his theory raised new questions about the nature of the phenomenal world, the importance and uniqueness of the human species, and the meaning of existence. The posing of these questions, for which no ready secular answers were found at a time when religion was losing credibility, added new impetus to the philosophical trend toward nihilism.

As Darwin maintained in his *Origin of the Species* (1857), the evolution of plants and animals was the result of a process of struggle for survival, what he called "natural selection." He argued that the survival of the fittest, based on chance and necessity, was the rule of nature. That is, chance genetic mutations in individual members of species were either useful or not in the struggle for survival within a given species and among different species. By a law of nature, those individuals, and species, with the useful characteristics were the ones that survived. Thus, there was a teleology at work in nature for Darwin, but it acted on chance elements and served no goal higher than biological survival and the operation of the evolutionary process.

In his view, the whole evolutionary process of nature had started with lower forms of life and had gradually produced higher forms. In his *Origin*, he stated that all animals were "descended from four or five progenitors" and remarked that analogy might even lead him to proclaim that all plants and animals were descended from one progenitor.[47] In his later work, *The Descent of Man* (1871), Darwin went even further and explicitly stated that humans and apes had a common ancestor.[48] To add empirical support to his theory of the evolutionary emergence of human beings, he pointed out the close resemblance between dog and human embryos and the occasional appearance of muscles in humans which they did not usually possess but that were common to mammals of the *Quadrumana* [primate] order.[49] In other words, Darwin believed the law of natural selection applied to humans as well as to animals; and he believed that chance mutations, as well as acquired characteristics, in some instances,[50] would be genetically transmitted by those who survived the struggle for existence and reproduced. Thus, over time, human nature was malleable; and the characteristics of human groups, or even the human race as a whole, were subject to change.

In his *Descent*, Darwin expressed his astonishment when he first saw, during his ocean voyage on the *Beagle*, the indigenous people of Tierra del Fuego and realized they resembled the ancestors of humans like himself. He interpreted the fact that "civilized" humans like himself had evolved over time from such aboriginal human types as a reason to hope "for a still higher [i.e., human] destiny in the distant future."[51] Today, however, there are some who want to bypass the evolutionary process entirely by producing immediate, predetermined changes in individual human beings by means of biotechnology.

Darwin's theory of natural selection, in any case, had a number of radical implications for Christian theology. In the first place, the idea of a dynamic, ongoing process of evolution involving chance mutations was incompatible with the account in Genesis of a separate and deliberate creation of all existing life forms in the beginning. And by showing the importance of chance mutations in the creation of new forms of life, Darwin provided new ammunition for those who argued that the universe was without higher purpose.

Second, if Darwin were correct in arguing human beings evolved from pre-existing, lower forms of life, then Adam and Eve, the first humans, according to Genesis, could not have existed when the world began. And if the biblical story of original sin and the Fall of man were false, then Christians would not need a messiah like Jesus Christ to rescue them from the wrath of a God who had condemned humanity in perpetuity to death as punishment for the transgression of its original members. In other words, the acceptance of Darwin's theory of evolution threatened to undermine the whole theological basis of Christianity.

In the light of such implications, it was not surprising that Samuel Wilberforce, an influential Anglican bishop of the period, condemned Darwin as a materialist and an atheist.[52] Nor was it difficult to see that when Nietzsche later proclaimed the death of God and the collapse of all higher values, the ideas of Darwin had influenced his thinking and helped prepare the ground for his declaration of nihilism.

In *Man's Place in Nature* (1863), Darwin's primary expounder in England, the biologist Thomas Huxley, was the first scientist to treat the human being explicitly as an animal and to state openly that humans did evolve from apes. Huxley pointed out that the physiological differences between some members of the primate order were greater than those between humans and apes. He also objected to the idea of a fall of humans from a higher to a lower condition because the fossil record showed the opposite—that is, that humans had evolved from a lower form, apes.[53] The effect of his view was to reinforce the moral and theological implications of Darwin's theory of evolution and to make a further contribution to the phenomenon of nihilism.

Comte and the positivists' promotion of science. Although religious belief and philosophical certainty were eroding, an increasing number of the dominant bourgeoisie of the nineteenth century lived by faith in the secular ideology of progress and believed that positive science, which validated its truth by the success of its practical (i.e., technological) applications, was the answer. This view not only owed a lot to the demonstrated successes of post-Baconian science, but it also was clearly in the line of the progressive tradition of the Enlightenment *philosophes*. Auguste Comte, who had been the secretary of the socialist thinker St. Simon in his earlier years, was one of the originators of sociology and the founder of positivism, a new philosophy based on scientific goals and methods.

In the decade of the 1850s, Comte published a number of works in which he not only asserted that he had determined the laws of progress, but stated that all knowledge, if it were based on a scientific approach, would become "positive." That is, the practical application of the results of scientific investigations would clearly demonstrate their theoretical validity. Furthermore, Comte asserted that a positive knowledge of social phenomena provided by his new science of society, sociology, would make it possible to orient knowledge as a whole correctly. The world would thus be transformed for the better; and humanity would be guided into its final form, without the violence and turmoil advocated by the Marxists.

Despite—or perhaps because of—its scientizing tendency, Comte's philosophy had an important sentimental dimension. Love, he said, was the motive behind all human endeavors and would provide the true moral center of his positivist world of the future. Women, he believed, were morally superior to men and represented the loving element in humanity. They were the ones, he said, who would ensure the triumph in the family of the positivist spirit in the world.[54] And as civilization progressed, the beneficial progress of chastity would lead to the phasing out of sexual reproduction and a situation where "birth would emanate from the woman alone."[55]

Comte even advocated a positive and secular version of Christianity in place of conventional religion. For the worship of God, he substituted the worship of humanity—the dead, living, and those yet unborn. He borrowed from Catholicism by including rituals and sacraments, a new calendar, secular baptism, and a new priesthood to teach the new dogmas, based on Comtean doctrine and the laws of science.[56] In other words, humans who accepted his ideas could have the millennium and religion, but without God and without Christ.

Bernard's ideal of the scientist as dedicated servant of humanity. Going hand in hand with the positivist belief that science was the only reliable kind of truth and the source of proper guidance for society itself was the idea that the scientist as a type was a noble, disinterested servant of humanity. This belief could be traced back to Francis Bacon, his description of Solomon's House, an idealized research center manned by dedicated scientists, and his millenarian vision of history. Even in the seventeenth century, however, Bacon had made it clear that he expected scientists to look after their own professional well-being and to serve the particular interests of the state.

In the nineteenth century, Claude Bernard, a Frenchman who founded experimental medicine, personally embodied the ideal of the modest and tireless scientific researcher who saw his moral obligation as limited to discovering the laws of phenomena for the benefit of society. According to Bernard, the scientist was to remain deliberately neutral on all philosophical questions and, when describing results, was to keep his personal involvement to a minimum. Furthermore, he believed that philosophy could not legitimately assign any limits to science, since the two realms were distinct.[57]

Bernard's strict separation of science and philosophy was prefigured, of course, by Bacon's idea that science should investigate the "how," not the "why," of the world and by the requirement that the scientist be objective. The problem today, however, is that many scientists and technologists refuse to accept any moral or social responsibility for the negative consequences of the application of their discoveries, while insisting they are fulfilling their role as dedicated agents of a greater social good. Scientists and technologists who do research that can be used for weapons of mass destruction or use biotechnology to create profitable crop varieties with potentially harmful long-term ecological consequences maintain, for example, that they are only "doing science" and reject any suggestion that their activities are primarily self-serving or that they should abandon their efforts because there is an obvious potential for harmful or immoral use.

Nietzsche's Rejection of All Existing Philosophical Alternatives and His Proclamation of Nihilism

The shortcomings and contradictions of positivism, as well as those of other main currents of nineteenth-century thought, be they scientific, philosophical, or religious, led ultimately to Nietzsche's critical response in its closing decades. In opposition to the ideas that history, art, science, reason, religion, or faith-based willing were the truth, Nietzsche asserted that there was no truth. He pointed out the "human-all-too-human" tendency to search for the truth, however, even when humans knew it didn't exist. Such a remark was not only an obvious reference to Kant's "as-if" approach to God, but also to Fichte, who proclaimed that a life of the striving of the will was ultimately a life based on faith. For Nietzsche, this tendency was an example of what he termed the "epistemology of nihilism."[58] What humans referred to as truths, he said, were only illusions, worn-out metaphors.[59]

This critique meant that as far as technology was concerned, the framework of higher values necessary for orienting it according to the classical Apollonian tradition no longer had any validity. In other words, the cultural horizon was clear for the philosophical rationale of a morally autonomous technology that appeared in the early decades of the next century.

THE FAMOUS "DEATH-OF-GOD" PROCLAMATION

In his first aphoristic work, *The Gay Science* (1882), Nietzsche proclaimed the "death of God" and the advent of nihilism. His announcement of this "tremendous event,"[60] however, was not without its antecedents. In *Faith and Knowledge* (1802), Hegel had written about "the feeling on which rests the religion of the modern period—the feeling God himself is dead."[61] Goethe had predicted nihilism in *Epochs of the Spirit* (1817), in which he described a coming age of prose, with a reversion to the primeval, the unholy, and the barbaric. The result, he warned, would be a period of strife, chaos, historical catastrophe, and the triumph of insipid mythologies.[62]

Heinrich Heine, a German-Jewish poet living in exile in Paris, had also anticipated Nietzsche with a "death of God" proclamation in his essay, *De l'Allemagne* (1834). In that work, Heine explained that Kant had shown the inability of reason to prove the existence of God and the futility of metaphysics. The vacant throne of religion, Heine said, was now "thought over" by philosophers. He predicted a period of philosophical wars; and when the "German thunder" came, he warned prophetically, there would be a crash like the world had never known.[63]

In a passage of Nietzsche's *Gay Science*, a madman lit a lantern in the marketplace and proclaimed to the crowd: "Whither is God? . . . I shall tell you.

We have killed him ... What did we do when we unchained this earth from its sun? ... Is there any up or down left?"[64] What this rejection of the Christian God actually signified, however, was a rejection of the entire supersensory world, the Platonic and Kantian realm of ideas and ideals. Meaning, as well, the death of metaphysics and a collapse of all higher values, it was, indeed, a proclamation of the arrival of nihilism.[65] It was an acknowledgement, in effect, that reason could neither reveal the structure of reality nor dictate moral truths. The madman's words about "unchaining the earth from its sun" were not only a reference to the unsettling effects of the Copernican revolution in astronomy for science, religion, and philosophy; they also alluded to Plato's analogy between knowledge of the Good and the sun, both of which Plato had said illuminated everything else.[66]

This declaration of the collapse of higher values played a crucial role in preparing the philosophical terrain for the emergence, after World War II, of a technological society in which technological goals and norms would prevail despite continued but hypocritical reference to these dead values. It likewise prepared the way at the end of the century for a technoculture in which ever more powerful and sophisticated technology would explicitly become, in both a theoretical and practical sense, the basis of culture.

THE CHARACTERIZATION OF ALL VALUES AS A RATIONALIZATION OF SELF-INTEREST

Concepts and values, Nietzsche argued in his philosophical critique, were nothing more than expressions of a subjective will to power, meaning they were only a rationalization of whatever served one's self-interest and had no claim to universal validity. Accordingly, he announced in *Beyond Good and Evil* (1886-7) that "psychology shall be recognized as the queen of the sciences." Behind all logic, he said, lay "psychological demands for the preservation of a certain way of life."[67] Furthermore, since values and concepts were only a rationalization of one's self-interest, there was "no limit to the way the world can be interpreted."[68] "The world," he declared, "does not exist as a world in itself; it is essentially a world of relationships; it has a different aspect from every point ..."[69] Each philosophy, therefore, was only the product of a particular perspective; and the conflict of perspectives in the world was inevitable.

THE FUTURE AGENDA FOR PHILOSOPHERS: A GENEALOGICAL FOCUS ON LANGUAGE

A corollary to the idea that all philosophies and values were nothing more than a rationalization serving the preservation of one's own interest, according to Nietzsche, was the fact that words were the raw material from which this false reality was constructed. To escape the prison of the false world we have created and make the world anew, he said, a genealogical analysis of language had to be made. Its purpose would be to expose the falsity of concepts and names such as "free will," "causality," and the idea of an "actor" independent of one's actions—all central to our psychological survival needs.[70]

Nietzsche's proclamation of the need to focus on language became his charge to the philosophers who succeeded him in the twentieth century. As we shall see, Heidegger honored Nietzsche's appeal to reject "knowledge" based on concepts and names; the analytical philosophers redefined philosophy in terms of the analysis of linguistic statements; and deconstructionists like Derrida attacked the logos, or rationality, of linguistic texts themselves.

THE ATTACK ON SOCRATES AND THE APOLLONIAN LEGACY OF MORAL LIMITS

To try to escape from the inevitable conflict of ideas and values, in Nietzsche's view, was to try to escape from life itself. Those who were strong, however, instinctively grasped within their own being, according to Nietzsche, that the truth was whatever was life-enhancing for them. They adopted an authentic will to power, he maintained, by openly affirming their own per-spective and recognizing that the world was whatever they could create from their own point of view.[71]

The problem in Nietzsche's view, however, was that European culture had incorporated as its guiding principle what he considered to be a "slave morality" of the weak, which he traced back to Socrates and Christ. Nietzsche praised classical Greek culture for knowing how to combine the irrational, instinctual aspect of human nature, represented by the earth god Dionysius, and the rational element that assigned limits, symbolized by the sky cult of Apollo. The Diony-sian factor was included symbolically, he said, in the satyr, a mythological creature that was half human and half goat; and the writers of Greek tragedy had wisely combined, he said, both the Dionysian and Apollonian elements in their works.[72]

Nietzsche derided Socrates, however, saying he was "the opponent of Dionysius" and was a "theoretical man," for whom thought was "capable of penetrating being" and even "correcting it."[73] In Nietzsche's view, the triumph of this repressive morality in Western culture represented the rule of the Apollonian principle to the exclusion of the Dionysian, leaving the human wounded in his soul. Such morality was an inversion of rank, he said, which amounted to "getting the eagle to think in terms of the lamb."[74] Its misguided purpose, he charged, was to enable the weak, who were more numerous, to prevail.

To comply with this kind of slave morality, said Nietzsche, the strong had to engage in a perverted, inward form of willing and repress actions which they were naturally, and therefore rightly, inclined to perform. The cultural result of this, in Nietzsche's view, was decadence. "The European disguises himself," he said in disgust, "in [this] morality because he has become a sick, sickly, crippled animal . . ."[75]

Nietzsche expressed a special contempt for Christianity, which he termed "a Platonism for the people" and associated with slave morality. The sick and decadent, he stated, meaning Christians, claim to have found truth in another world.[76] He scoffed at the Bible's promise of heaven to those who had become "little children." "We have no wish whatsoever," he said, "to enter into the kingdom of heaven: we have become men, so we want the earth."[77] This statement reflected his rejection of any transcendence and his belief that everything depended for humans on their lives in the phenomenal world.

This attack by Nietzsche on limits had a decisive influence on Western thought in the twentieth century. Along with his glorification of will, it set the stage for the reactionary modernists' glorification in the Weimar period of technology as the means for realization of the political and cultural destiny of the German people.

SCIENCE DISPARAGED AS A FALSE AND MEANINGLESS PATH

Nietzsche extended his critique of European culture to the objectifying perspective, so often praised, of the scientist. For him, it was only an example of the crippled will to power of the weak and the ordinary. Objectivity reduces a man, he said, to a "mirror . . . an instrument" with "no goal, no conclusion."[78] The "scientific man," he declared, ". . . is not noble" and displays "an unwillingness to take risks."[79]

Nietzsche likewise rejected the validity of the natural sciences, where he found only "meaninglessness, casualism, and mechanism."[80] He explained that a logical concept like causation had been invented by humans in an attempt to control the external world by making it appear as a sequence of logically classifiable and recurring events. Causes were merely added on to events, he stated, concluding, "There are neither causes nor effects."[81] The scientist's attempt to reduce everything to laws and uniformities, he said, derived from "a plebeian antagonism to everything privileged and aristocratic."[82] Such remarks applied, of course, not only to scientists, but to engineers and technologists as well.

Despite his objections to science, however, Darwin's scientific theory of evolution had a profound influence on the thinking of Nietzsche. According to R. J. Hollingsdale, it made the most telling contribution to Nietzsche's perception that European culture had reached a point of nihilism. The theory of natural selection, which Nietzsche interpreted as giving the universe over to chance, was proof for him that there was no need for the hypothesis of a directing agency. It meant that humans were living in a universe without any higher meaning or purpose.[83]

A FAILURE TO SEE ANY MEANING OR GOAL IN HISTORY

Like Schopenhauer, and in contrast to philosophers like Hegel, Nietzsche ridiculed the idea that history was a bearer of any truth or meaning. In concurring with the pre-Socratic, mechanistic conception of the world as a place without meaning where everything was in motion, Nietzsche asserted that the only historical possibility was what he termed "eternal recurrence"—that is, everything coming back to the same.[84] It was existence, he argued, the world of phenomena, that contained the totality of possibilities;[85] but there was no linear teleology of historical development.

The goal of history did not lie for Nietzsche at its end but would be found, rather, in its higher individual specimens—that is, representatives of a new aristocracy that he summoned to appear.[86] The overman (translated from the German *Übermensch*) would be the one who practiced what Nietzsche called master morality and who would live totally in the present. He would be his own standard and would not base his existence on a sense of future time when things would be better.[87] He would be willing to accept the world as it was, with all its joys and sorrows, and without any possibility of redemption.

Nietzsche characterized historical explanations as "fatalism, Darwinism; the final attempts to read reason and divinity into" events and "sentimentality in the face of the past." Will, or the force within, he argued, was more important than the influence of milieu and external causes.[88] This rejection of historicism carried with it, of course, an implied rejection of all the millenarian expectations for technology that were part and parcel of the ideology of progress. It also signified a rejection of the Marxist revolutionary project and the communist ideal, whose egalitarianism was as repugnant to him as its historicism.

REPUDIATION OF THE IDEA THAT ART IS THE ANSWER

When he was young and had occupied a professorial chair in philosophy at the University of Basel, Nietzsche had embraced the idea that art was the answer to the crisis of nihilism in European culture. This was one of the reasons why the historian Jacob Burckhardt, a colleague at the university, referred to Schopenhauer in conversation with Nietzsche as "our philosopher."[89] Nietzsche even believed for a time that Richard Wagner, the German genius and composer, would be the artist who would rescue European civilization from its crisis. He would accomplish this, said Nietzsche, by uniting in his music archaic myth, which contained the repressed but necessary Dionysian element of culture, and modernity. Wagner, who had met Schopenhauer in 1854, shared Schopenhauer's belief in art and incorporated in his opera *Tristan and Isolde* the idea that redemption would come only through escape from the world.[90]

Eventually, however, Nietzsche abandoned his faith in art and terminated his relations with Wagner, who at one point had been a father figure for him. He ultimately qualified "art for art's sake" as "aesthetic pessimism" and associated it, too, with nihilism. The artist, Nietzsche said, was an intermediate type brought on by "increasing civilization," a variety of human "restrained from crime by weakness of will and social timidity, and not yet ripe for the madhouse . . ."[91]

BELIEF IN THE SUPERIORITY OF THE VALUES OF *KULTUR* OVER *ZIVILISATION*

Making it a special point, as had Hegel before him, to distinguish the vital and authentic category of the values of *Kultur* from their corrupted variant, the values of *Zivilisation*, Nietzsche identified the latter with France, Rousseau, and the Enlightenment. He referred to the "feeble-optimistic eighteenth century" and criticized its promotion of freedom, community spirit, pity for the underdog, and progress.[92] "The high points of culture and civilization," Nietzsche wrote, do not coincide:

> [O]ne should not be deceived about the abysmal antagonism of culture and civilization. The great moments of culture were always, morally speaking, times of corruption; and conversely, the periods when the taming of the human animal [i.e., by civilization] was desired and enforced were times of intolerance against the boldest and most spiritual natures.[93]

Although he agreed with Rousseau's rejection of the Enlightenment theory of progress, Nietzsche otherwise singled out Rousseau for special contempt. He condemned, for example, Rousseau's preoccupation with women, suffering, and feeling—what he called "sensualism in matters of the spirit." Furthermore, he maintained that Rousseau had made the mistake of arguing that justice could be based on a sentimentalized view of nature. Nietzsche's response was to quote Voltaire, who had said the state of nature was "terrible" and that humans were "beasts of prey."[94]

His preference for *Kultur,* in any case, paved the way for the reactionary modernist philosophers of the Weimar Republic, who emphasized traditional German values (i.e., the "reactionary" element in their philosophy) and associated them with will and technology (i.e., the "modern" element) to produce an aggressive and nationalistic ideological program which the National Socialists subsequently adopted and attempted to put into practice.

A DUBIOUS CALL FOR THE WILL TO POWER OF THE OVERMAN TO OVERCOME NIHILISM

Although he clearly recognized the phenomenon of nihilism as evidence of a profound crisis in European civilization, Nietzsche believed that it presented an opportunity for a positive transformation. The death of God, he argued, had shattered all equality among humans, who up to that point had seen themselves as equal before God.[95] Henceforth, he said, humans were free to create their own values—and truths—and the door was open for the overman, who would assert his superiority and practice an authentic will to power. The goal was a new form of human, who would embody knowledge and make it instinctive, going beyond the falsity of words and the illusions of truth.[96]

Predictably, however, Nietzsche's philosophy was not a solution to nihilism; and his thought ultimately served only to deepen the crisis in European thought and culture. His call for the emergence of a new aristocracy of Overmen who would practice master morality and create a better world for all was never really answered. Rather, a generation after his death and at the urging of his sister, whom Nietzsche reproached for marrying an anti-Semite, the Nazi regime made Nietzsche its official philosopher. To do so, however, its ideologues had to expurgate from his works some important elements that were incompatible with their own ideology,[97] such as his contempt for German nationalism and his disdain for socialism.

His concept of the will to power, however, was too easily interpreted by them to mean domination of the weak by the strong, not the more profound transformation of values for which Nietzsche had called. Furthermore, although he considered technology a reactive force, the will to power principle easily lent itself to the belief, dear to the reactionary modernists and Nazis, that technology, a constantly escalating means to power, was a phenomenon which the German people could place all their confidence and see all their dreams fulfilled.

The Theory and Practice of Technology in the Nineteenth Century

CONSEQUENCES OF THE BREAKING OF PRACTICAL LIMITS BY INDUSTRIAL TECHNOLOGY

The modern Industrial Revolution, which had begun in the eighteenth century, gained additional momentum in the nineteenth century and brought radical changes to society, production, and the natural environment. On the level of practice, these changes demonstrated more clearly than ever that the process of technological development was not subject to predetermined limits and that the effects of specific technological innovations were so broad and complex that technology could not be accurately defined by the classical Greek formula of neutral means susceptible to either good or bad application as determined by external values.

Production was now concentrated in the factory, where the new machinery was driven by nonhuman power, and the new division of labor diminished significantly the level of skill needed by the worker and condemned him to the performance of dull, repetitive tasks. At the same time, steam-powered machinery magnified the productive capacities of the work processes far beyond all previous limits.

According to José Ortega y Gasset, when humans saw the new, power-driven machinery functioning independently of the worker, whose task had been reduced to "aiding and supplementing" it, they began to realize that technological operations had become a phenomenon distinct from ordinary human activity and not subject to the same limitations. They realized, Ortega asserted, that henceforth there were no a priori limits on what their machines would be able to do.[98]

The human dimension of negative effects: dehumanization. What Marx included in the category of "productive forces," and what today would include the organization of work and the software and procedures employed by the worker, also underwent a fundamental transformation during the Industrial Revolution. Babbage's *Treatise of Machinery and Manufactures* (1832) was an important theoretical rationale for these changes. His emphasis on the difference between making and manufacturing and between tools and machines reflected the transfer of production from the artisan's home or workshop to the factory where the proletariat toiled. Babbage's goal was to produce the perfect object for the lowest price, and to accomplish this he advocated reducing not only the time of a given production, but also the time required for learning the necessary operations. And the more work tasks were broken up into discrete operations, he declared, the more it would be possible to control precisely the strength and skill necessary for them.[99]

These changes signified the dehumanization of human labor, and the response of workers who had been skilled craftsmen in the workshop or home was predictable. In 1811, English weavers smashed the mechanical textile looms that deskilled the work they had to perform in the factory. The new machines made it possible for employers to automate and to hire unskilled women and children, bringing existing workers face to face with the prospect of unemployment and the threat of starvation. In 1812, the British government called out 24,000 troops and local militia to suppress the so-called Luddite rebels (they operated under the banner of a mythical General Ludd), some of whom were sent to the gallows. Although "Luddite" unfairly became a derogatory term for opponents of technology, Iain Boal reminds us that the Luddites were not mindless wreckers of machinery but artisans fighting against the dehumanization of their labor and the elimination of their livelihood.[100]

The ecological dimension of negative effects: pollution and the escalating assault on natural resources. The ecological contradictions of the Industrial Revolution were aggravated in the nineteenth century, as the air, water, and soil of the natural environment were subjected to more extensive and intrusive technological operations. Not only did black smoke from factory chimneys fill the air, but there were other forms of atmospheric and waterborne wastes, such as dust, fumes, and effluvia, that emerged from soda factories, ammonia works, cement plants, and gas facilities.[101] Streams were polluted with untreated sewage from dense urban concentrations, toxic chemicals from dye houses and bleach yards, discharges from boilers, and the waste products of mining operations.

More efficient technologies for extracting and smelting metals and ores served to create an increasing demand and accelerated consumption of these materials. Aided by the new, nonhuman energy sources that could be mobilized for processing and transportation, and enhanced by the increasing consumption demands of a growing world population, the assault on the world's raw materials escalated to levels without historical precedent.

DEVELOPMENTS IN THE THEORY OF TECHNOLOGY

Despite the human and ecological contradictions of the new industrial technology, a number of the most important philosophers of the century—Hegel, Marx, Engels, and Comte, all of whom were in the historicist tradition inherited from the Judeo-Christian theologians—assigned to technology a crucial and positive role in the accomplishment of human destiny. Schopenhauer and Nietzsche, who were not historicists, significantly did not attach any positive importance to technology. Their emphasis on will, however, eventually served the ideological promotion of technology. Early in the twentieth century, the same reactionary modernist and National Socialist thinkers who glorified will praised technology as a means for domination of both humans and nature.

For Hegel, the history of consciousness began with **techne.** When Hegel said that the history of consciousness, which for him *was* human history, began with labor, he included by implication *techne* in his formulation. In his text on "master and slave" in the *Phenomenology of the Spirit* (1807), he explained how humans first put some of their intelligence—that is, Mind—into nature by fashioning matter. And if humans put Mind into nature by means of their labor, there had to be a method—even for such primitive operations as breaking, scraping, or gathering—and very likely recourse to the use of implements, although very primitive at the outset. Either of these rational aspects of labor, in any case, signified *techne.*

Up to this point in time, according to Hegel, Reason had merely "slumbered" in nature. With the labor of the slave, however, Hegel said it "awakened" to self-consciousness and began its process of slowly unfolding in world history.[102] Thus labor—and *techne*, in Hegel's schema—enabled humans to negate their animality, break with nature, and render their own history possible.[103] There was clearly a Hegelian resonance to Ortega y Gasset's formula, pronounced in his important lectures on technology in Buenos Aires in the next century: "Man begins where technology begins."[104]

The kind of Mind that went into nature by means of work/*techne* as related in the *Phenomenology*, however, was what Kant had termed its understanding (*Verstand*). This type of thinking involved a subject-object mode of consciousness and produced empirical knowledge, abstracted from a greater whole and inferior to the higher, synthetic truths which, according to Hegel, were the product of higher reason, or *Vernuft*. By his criteria, therefore, technology alone could never pretend to offer any higher truth or constitute a value or goal in itself.

For Comte, the engineer, master of technology, would spearhead human progress. The earliest writings of Comte preceded those of Marx and Engels by a generation, but his doctrine of positivism was based on a similar quasi-religious faith in a historical resolution of the problems of humanity and a Baconian belief that applied science (i.e., technology) was to be the means by which human destiny would be realized. He had been a student at the French *École Polytechnique*, founded at the end of the eighteenth century and destined to become the world's premier professional engineering school. Not surprisingly, Comte assigned to the engineer a crucial role in the realization of human progress. In his view, engineers, who combined scientific theory and technological practice, would make the synthesis of science and industry upon which a new social order would be based. They would be, in effect, the new priesthood in his perfected, positivist society.[105]

In Noble's words, Comte was the "true herald of the engineer."[106] His writings signaled the entry of this new professional class, the possessors and practitioners par excellence of technology, into the forefront of the modern world. Some of its members, not unlike Bernard's scientific elite, began to see themselves as having an exalted social role. By the twentieth century, reactionary modernist philosophers and National Socialist ideologues likewise assigned a lead role to engineers, although in their case it was to provide the practical means by which the crisis facing Germany and European culture would be solved.

For Marx and Engels, the level of production technology determines the nature of culture and values. Marx's philosophy, in contrast to thinkers like Comte and Hegel, explained that technology had played a key role throughout history. He also differed from Hegel in characterizing a human being as a *homo faber*, or maker of things, not primarily as a thinking being. Despite these and other important dissimilarities, however, Hegel's section on master and slave in his *Phenomenology* had a decisive influence on Marx.

Hegel's assertion that it was the laboring slave, rather than the contemplative master, who enjoyed an authentically human relation with nature was reflected in Marx's assertion that the most fundamental thing a member of the human species did was to produce the goods needed for sustenance. For Marx, therefore, the worker had a true form of species existence, but the bourgeois capitalist, who performed no physical labor, did not. Similarly, Marx and

Engels's idea of class conflict, originating with the master-slave dichotomy of tribal society and continuing within the context of each higher stage of development of production technology and its exploitive division of labor, also bore the influence of Hegel's "master and slave" text.

Marx and Engels's view of technology implicitly included acceptance of the theodicy of Turgot and Condorcet's *laissez-innover*—that is, the idea that technological advances would inevitably benefit society as a whole. Thus, these thinkers all believed that technological change was not an end in itself but a guaranteed means to a good end. According to the Marxist philosophy of history, the level of the productive forces in the base of the social system (technology—machines, tools, equipment, methods and organization of production) at any particular stage in history determined the nature of the productive relations (class relations, property ownership, and distribution of goods), also in the base, as well as the elements of the superstructure (the type of government and the character of the thought system, morality, law, the family, etc.) When humans made a significant advance in the technology of production, Marx and Engels declared, all other aspects of life and culture would necessarily change as well, moving humanity closer with each transformation to the eventual goal of communism.

Higher values, as expressed by laws and morality, were thus an expression of material conditions, which were determined by the level of technology. In the strictly Marxist sense, therefore, technology could never have been accurately characterized in Apollonian terms as neutral means oriented by higher values, since it was always *the* element of civilization which determined everything else, including values. As Langdon Winner put it, technologies were for Marx "forms of life" whose cultural, social, and ecological effects always were greater than their immediate purpose and use as means.[107] Certainly, by the time the Industrial Revolution had gotten underway, the realities of everyday life appeared to confirm the validity of this conception.

For Marx and Engels, the negative aspects of industrial technology are not inherent but are due to capitalism. Marx gave a famous description of the dehumanizing effects of the technology of production during the capitalist phase. The worker, he said, was reduced to nothing more than an "appendage of the machine"; and the work process in the factory, he added, had greater importance than the human being who was involved in it.[108] He explained how the division of labor transformed it into boring, repetitive tasks, making possible the replacement of skilled workers by unskilled ones, including women and children who were paid lower wages. Furthermore, the physical and mental aspects of work were separated and performed by different individuals, and specialization was rife. The result, according to Marx, was that the worker, instead of feeling free in his work, which was his most essentially human activity, felt free only in his leisure time, in his animal functions of sleeping, eating, copulating, and so on.[109]

Steam power, advanced machinery, efficient and repetitive processes, and the factory system, all technological aspects of the rationalization of the means of production during the capitalist phase, were necessary in Marx and Engels's view to secure a transition to communism. That is, they were essential for producing the superabundance of goods that would enable each person to receive goods according to the communist ideal of need rather than the capitalist standard of wealth. Industrial technology would likewise secure a significant increase in human freedom from nature, provide the material basis for ending class conflict, eliminate the need for the traditional coercive state, and bring to a close the dialectic of history itself.[110] The problem was, however, what would become of the dehumanizing aspects of work associated with this same industrial technology when the communist utopia was realized?

Marx insisted that with the victory of communism, the specialization of industrial capitalism would end, the distinction between mental and physical labor would "vanish," work would become meaningful again, and humans would be free to pursue their all-around development.[111] In making these pronouncements, however, he abandoned his principle that the level of technology determined everything else in society and culture. That is, he reversed the causal relation between the superstructure of a social system and the productive forces in its technological base, arguing that the establishment of communism would eliminate the disadvantages of industrial production.

This theoretical inconsistency carried with it idea that the contradictions of modern industrial technology were not inherent in the technology itself. This is one of the most problematical aspects of Marxist theory, which in many important respects continues to provide a valid explanation and useful critique of capitalist society. It prevents doctrinaire Marxists, furthermore, from being open to a critique of modern technology, since they believe its shortcomings derive from capitalism rather than the technology itself. It is significant that from a later historical perspective, Jacques Ellul forcefully made the case in his *Technological Society* (1955) that the most important shortcomings of modern technology derived from its own methods and goals and could not be eliminated by changing the social system.

Another reason for Marx and Engels's lack of a more critical perspective on industrial technology as such no doubt derived from the fact that they were thoroughly steeped in the progressive tradition of the Enlightenment and the theodicy of *laissez-innover*, according to which technological development always turned out to be a blessing.

For Marx and Engels, technological advance could be successfully combined with cultural regression in communist society. Rousseau, as we have seen, did not believe humans would find salvation through technology. He not only argued that it had triggered an initial chain of historical events that had led to human misfortune, but he asserted that it was a continuing source of human misfortune. His formula in the first *Discourse* for regaining happiness and goodness suggested cultural and social regression to a simpler life, close to nature. Marx and Engels, on the other hand, were in the mainstream of the Enlightenment *progressiste* tradition and considered technological advance necessary for establishing conditions under which humanity would be perfected. For them, in other words, technology was not part of the problem and would be a basis for the solution.

Their formula for healing the ills of society, however, was both an advance and a regression in the historical sense. Like the new-old synthesis of Bacon and the Puritan millenarians of the seventeenth century, who had combined technological advances based on a new science with recovery of human characteristics from a distant past before the Fall, Marx and Engels combined in their utopian vision technological advance with a return to the communal social arrangements and freer erotic relations of a distant barbarian past. Thus, Engels explained in his famous essay, *The Origin of the Family, Private Property, and the State* (1884), that the communistic property arrangements, participatory decision-making, classless social relations, and freer sexual morality of the barbarian period provided the model upon which the communist society of the future was to be based.

That it might be difficult to combine these historically distanced elements in a viable social entity was suggested by the reality of the nineteenth-century American commune experiments which Engels confidently cited in an 1845 article as empirical "proof" that his and Marx's form of communism was feasible. In fact, Engels had no direct personal experience with these communal experiments, and he based his blueprint in the *Origin* on an 1877 anthropological study by Lewis Morgan of primitive communal groups called *gens*.

The truth about the American communes was that none was a combination of the industrial production, communistic property arrangements, and the liberated sexuality that Marx and Engels said would characterize the coming communist society. Agrarian and craft elements actually prevailed at the Brook Farm commune and in the communal settlements of Shakers, Rappites, and Separatists to which Engels referred;[112] and none had industrial operations of scale or utilized the advanced production technologies of the time. Furthermore, the latter three groups had a religious basis for cohesion, which was anathema to Marx and Engels's secular ideal; and the Shakers, who enjoyed perhaps the greatest measure of material success, forbade marriage and practiced celibacy, which the Rappites also promoted, in contrast to Marx and Engels's goal of freer erotic arrangements.

The greatest problem, indeed, was in the proposed sexual arrangements under communism. Engels asserted that in capitalist society, true love was only possible for members of the proletariat, since their conjugal unions were not based on property relations.[113] Accordingly, he wrote that true love for all would be possible only after private property, the basis of male dominance and the cause of the falsification of love within the traditional family, was abolished in a communist society.

According to Engels, individual sex love, rooted in reciprocal love and male-female equality, would become the new erotic standard in a communist society. A "gradual rise of more unrestrained sexual intercourse" would occur, he added, since public childcare would reduce the vulnerability of women by providing for children born out of wedlock. Sexual relations would be considered legitimate if they were based on love, and the individual would no longer care about the taboos and proscriptions dictated by bourgeois sexual morality.[114]

Given the physical strain of factory labor and the dehumanizing character of the power-driven machines and efficient processes upon which industrial production relied, however, a crucial question was whether they could realistically be combined with the vital and freer eros that Marx and Engels advocated. Fourier, the French anarchist-socialist who published his *Grand Traité* in 1822 and wanted to combine in his utopian communities (*phalanxes*) the assurance of a minimum material standard of living with humane, libidinized working conditions and the guarantee of a minimum level of sexual gratification for all, essentially answered this question in the negative. To create a social organism where both work and love would be liberated, Fourier pointedly stated that the members of his rural-based communes should devote no more than one quarter of their labor time to manufacturing.[115] Furthermore, he declared that care of animals, agriculture, and kitchen work, in a descending order of suitability, were all to be preferred to manufacturing because they were more conducive to the realization of his work- and pleasure-related goals. Thus, although Marx and Engels considered advanced industrial technology and liberating social and sexual arrangements to be compatible, Fourier believed they were not.

Russell Means, a twentieth-century Native American leader and a radical political thinker in his own right, took issue with the general proposition that the Marxist economic model was compatible with the spiritual, nature-respecting cultures of tribal peoples similar to those of the barbarian period. In Means's view, Marxism was a materialist doctrine that obeyed an ethic of efficiency. "Hegel and Marx," he stated derisively, "were the heirs to the thinking of Newton, Descartes, Locke, and Smith."[116] He accused Marxists of putting the emphasis on "gaining," although "being" was more important to tribal peoples. For them, Means declared, living according to the principles of Marxism would only be cultural genocide.[117]

For Goethe, a world of technology would be prosaic. Goethe and Nietzsche, although their writings on the subject were approximately fifty years apart, were nineteenth-century thinkers who did not share the positive view of technology that Comte, Marx, and Engels had inherited from the Christian millenarians, Bacon, and the progressivist *philosophes* of the Enlightenment. In *William Meister's Wanderjahre* (1829), Goethe's protagonist looked through an astronomer's telescope and commented on the disquieting discrepancy between the increasing level of knowledge of external nature furnished by technological devices and humans' limited inner level of discernment.[118] Goethe, in effect, believed there had to be a balance between what a person was and what he had the power to do. Accordingly, he warned prophetically about technological hubris. "The promise of Heaven for the poor in spirit is understood," he said, "to mean that on earth, at least, they should be educated into clever people able to manipulate and let loose the technological installations of Hell."[119]

He also warned of the inability of art to distill real poetry from a world impoverished by scientific abstraction and a preoccupation with utility. "A world empty of meaning," he said, in terms which seem to have a contemporary relevance, "seeks to escape from the infinite boredom of its meaninglessness by the magic of words without flesh, and forms without content."[120]

For Nietzsche, science is a "reactive force." Nietzsche praised autonomous science, or "the emancipation of science from moral and religious purposes," as a "very good sign."[121] He praised it, however, not for its material rewards, but because he saw it as a symptom of the cultural nihilism which he believed presented an opportunity for a radical and positive transformation of values. He also credited science with accomplishing the beneficial task of undermining Christianity.[122] Similarly, he said that the "transformation of humans into machines" that was taking place in contemporary culture was a precondition for the emergence of a higher and necessary form of human being, the overman.[123]

Nietzsche's overall assessment of science, however, was clearly negative. This judgment applied equally to technology, since after Bacon, science was organically linked to its technological applications. In effect, Nietzsche considered science and technology to be reactive forces, the expression of a false will to power seen from the perspective of the slave. In effect, from Nietzsche's perspective, neither of the *techne*s, art nor technology, could provide a valid response to nihilism.

Science, said Nietzsche, applying his psychological interpretation of thought, was based on fear——"fear of wild animals . . . including the animal he [the human] harbors inside himself . . ."[124] The idea that science had its equivalent and measure in human thinking, he said, robbed existence of its fundamental ambiguity. Echoing Goethe, Nietzsche stated that "an essentially mechanistic world would be a meaningless world."[125]

The scientific, positivistic attitude revealed the same need for certainty as Christianity, he said; and both reflected the same "pessimistic gloom."[126] Furthermore, the promotion of science (and by implication, its technological applications) was another expression for Nietzsche of the false values of *Zivilisation*. The reactionary modernists and National Socialists who followed him in early twentieth-century Germany, however, took the fatal step of transferring technology to the authentic realm of *Kultur*, whose values they exalted. In doing so, they presented it as a morally autonomous and powerful thing in itself, linked with German soul and a nihilistic will to power.

Anticipation by Realist Artists and Writers of a Twentieth-Century Theory of Morally Autonomous Technology

THE SHIFT OF EMPHASIS IN PAINTING FROM SUBJECT MATTER TO TECHNIQUE

The philosophical rationale for the *techne* of art for art's sake, which was developed in the nineteenth century, was applied to technology in the twentieth century by reactionary modernist philosophers and National Socialist ideologues to give it a morally autonomous basis. As we have already indicated, Kant had cleared a path for this development when he separated art from ethics, science, and metaphysics and designated it, essentially, as an autonomous activity. Schopenhauer's assertion that art provided a way to transcend the unhappiness and meaninglessness of existence was a further step leading to the idea of art as an end in itself.

Some of the most important innovators of a new, morally autonomous aesthetic of art for art's sake in the nineteenth century were painters. For them, the form of a work of art was more important than its subject matter, which could be anything. In 1834, for example, a critic praised "The Woman of Algiers," a canvas by the leading Romantic painter in France, Eugène Delacroix, as a work of art "without a subject." His words referred to the fact that the subject matter of the painting, rendered in heavy, free brush strokes and in striking colors, was secondary in importance to its visual qualities.[127]

This shift of emphasis from content and meaning to form and technique was adopted by painters of the so-called Realist school. Gustave Courbet's famous canvas, "The Burial at Ornans" (1850), shocked the public because death and religion, instead of inspiring noble thoughts, were used as a pretext for presenting local peasant types in ridiculous costumes. Even more significant, however, was the fact that the aggressive presence of the individuals in the scene was paralleled by an aggressive presence of paint.[128] By building up the paint on the canvas, Courbet insisted on the technique of the painted surface as no one had ever done previously. At the same time, he conveyed only emotional and moral indifference to his subject matter.[129] He thus shifted the focus of his work from content to the style and technique of visual representation.

The impressionist painters (1874-88) who followed him continued in the same vein. Monet, for example, used broad and broken brush strokes to render luminosity in a stylized way. Van Gogh, a postimpressionist whose paintings included a rendering of an old pair of shoes, built up the paint on his canvasses with his palette knife to give them physical relief; and he often deliberately used violent and shocking colors. These innovations served to affirm that the artist had a right to make a work of art out of anything and that technique was more important than moral, historical, or religious content.

THE CLAIM THAT A BEAUTIFUL STYLE WAS SUFFICIENT MORALITY FOR A LITERARY WORK

In literature, Flaubert and Baudelaire were representative of the Realist writers who treated art as a higher value and created a new, morally autonomous aesthetic. Flaubert's equivalent to Delacroix's idea of a "painting without a subject" was his idea of a book in which the style would be beautiful and the subject matter of secondary importance. In 1852, Flaubert wrote to Louise Colet: "What I would like to do is a book about nothing, a book without any exterior tie, which would hold together by the inner force of its style . . ."[130]

This new aesthetic amounted to a rejection of conventional morality in favor of a new supreme value, art. It was not only a response to the nihilism of the time, but in another sense it was an expression of it. Although Flaubert related that he found his characters in *Madame Bovary* "repulsive," he avoided any moralizing comment about them.[131] The morality of this new Realist aesthetic, rather, was in the beauty of the style itself. Morality, in essence, was being aestheticized.

The Realist writer chose his subject matter in complete freedom. For example, one poem that appeared in Baudelaire's famous collection, *Flowers of Evil*, was "Une Charogne" (A Carcass), a shocking but typical demonstration of beautiful form applied to content of the most arbitrary and unsavory character. In that strange piece, the poet addressed his beloved, declaring that one day she, too, would be like a putrefied, dead animal they had encountered lying rotting in the road. After describing the carcass in beautiful verse, he concluded with an appeal to his beloved to tell the vermin which one day would "eat her with kisses" that he, nevertheless, would retain the "form and divine essence" of his "decomposed loved one."[132] Baudelaire thus shockingly contrived to exalt the spiritual and the absolute in beautiful verse while treating a degraded and physically repugnant subject matter.

Baudelaire was also important for his visionary recognition that continual change, impermanence, and experimentation would be an essential characteristic of the new art. "Modernity," he wrote, "is the transient, the fleeting, the contingent; it is one-half of art, the other being the eternal and the immoveable."[133] He referred to this relative, circumstantial element as that "which we like to call contemporaneity, fashion . . ."[134] He understood, in other words, that with the art-for-art's-sake revolution, experimentation in form was becoming the new, actual content of art. Henceforth, the process of change—that is, continual innovation, primarily in technique—would be the essence of modern art. This meant that the modernist avant-garde would have an importance without precedent in the world of art.

The response of conventional society's guardians of morality to the aestheticized ethics and emotional detachment of this new kind of art was predictable. Both Flaubert and Baudelaire were put on criminal trial in 1857 by French authorities for offending public and religious morality. At the first trial, which followed publication of *Madame Bovary*, the public prosecutor condemned Flaubert and Realist literature, not for depicting such things as "hatred, vengeance, and love," but for depicting them coldly and "without restraint, without measure." Although the trial ended in acquittal, the judge admonished Flaubert for not recognizing that "there were limits which even the highest literature should not transgress."[135]

The decision of the court in the case of Baudelaire's *Flowers* was even less favorable to the new aesthetic of beautiful style without moral content. A number of critics helped prepare the ground for the trial's negative conclusion. Although Thierry published a review of it in which he praised the book as a work of fierce realism, written in a noble style,[136] a detractor wrote in the pages of the *Journal of Brussels*: "Nothing could give you the idea of the heap of filth and horror contained in the book."[137] And despite Baudelaire's response that the sadness and despair of his book were sufficient morality and that the work expressed a "terror and horror of vice," the court found him guilty of offending public morality by creating a sordid realism that led to unhealthy stimulation of the senses.[138] He was fined fifty francs, and six poems were ordered removed from any future edition.

From a modernist perspective, the court's pronouncements may appear to be excessive. But in another sense, the public authorities who ordered them, like the inquisitors who had prosecuted Galileo two centuries earlier for his innovations in natural philosophy, correctly sensed that fundamental, long-held values were being radically undermined. The avant-garde practitioners of the *techne* of fine art, like the revolutionary scientists of the Renaissance before them, were seeking a new conception of truth and were casting off limitations imposed by the dominant institutions and higher values of the society in which they were living.

The philosophical rationale for the other kind of *techne* as defined by the ancient Greeks, which today we call technology, was destined to undergo in the twentieth century a transformation similar to that which had affected the *techne* of art in the nineteenth. This transformation would likewise signify a transcendence of existing cultural standards and a claim of moral autonomy. The difference, however, was that the influence of art was primarily symbolic and upon the sensibilities of a cultivated segment of the population, whereas the glorification of technology and its severance from external values contributed to the multiplication of the destructive effects of war, the heightening of the threat to nature's integrity and survivability, and the radical alteration of the everyday conditions of life and work of countless human beings.

Notes to Chapter 5

1. Fredrick Coppleston, *A History of Philosophy: Fichte to Nietzsche*, vol. 7 (Westminster, Md.: The Newman Press, 1965), 207.

2. Sabine, 634.

3. Coppleston, 213.

4. Sabine, 640-41.

5. Ibid.

6. Ibid., 666.

7. Maurice Mandlebaum, *History, Man, and Reason: A Study in Nineteenth Century Thought* (Baltimore; Johns Hopkins Press, 1975, 281.

8. Sabine, 627.

9. Ibid.

10. Sabine, 280-1.

11. Karl Marx, "The German Ideology," in *The Marx-Engels Reader*, 2d ed. (New York: W.W. Norton & Co., 1978), 172.

12. Friedrich Engels, "Socialism, Utopian and Scientific," in *Essential Works of Marxism*, 2d ed. (New York: Bantam Books, 1963), 80.

13. William T. Bluhm, *Theories of the Political System* (Englewood Cliffs: Prentice Hall, Inc., 1965), 421.

14. Marx, "The Communist Manifesto," in *Great Political Thinkers*, 3d ed (New York: Holt, Rinehart and Winston, 1960), 700.

15. Engels, *Essential Works of Marxism*, 80-81.

16. Augustine, 854.

17. Coppleston, 7, 51, 53-54.

18. Johann G. Fichte, *The Vocation of Man* (Chicago: Open Court Publishing Co., 1910), 5556. William Smith in *Johann Gottlieb Fichte's Popular Works, with a Memoir by William Smith, L.L.S.* (London: Trubner and Co., 1873), 67.

19. *Vocation*, 109.

20. Ibid., 136.

21. Ibid., 109-11.

22. Smith. 62.

23. Fichte, Vocation, 136-38, 140.

24. George Santayana, *The German Mind* (New York: Thomas Crowell, 1968), 67.

25. Coppleston, 17, 28.

26. Edman, Irwin, editor. introduction by editor, *The Philosophy of Schopenhauer* (New York: Modern Library, 1928), 152, 266.

27. Ibid., 149.

28. Coppleston, 273.

29. Edman, 64, 71.

30. Ibid., 248.

31. Ibid., 249.

32. Ibid., 254, 256.

33. Ibid., 223.

34. Ibid., 155.

35. Ibid., 161.

36. Johann Wolfgang von Goethe, *Faust: A Tragedy* (New York: The Modern Library, 1967), 60.

37. Fedor Dostoevsky, "Notes from the Underground," in *The Short Novels of Dostoevsky* (New York: Dial Press, 1945), 150-51.

38. Ibid., 147, 149, 151.

39. Ibid., 150.

40. Ibid., 149.

41. Samuel Florman, "In Praise of Technology," *Harpers*, Nov. 1975, 70.

42. Richard Terdiman, *The Dialectic of Isolation* (New Haven: Yale University Press, 1967), 63.

43. Enid Starkie, *Flaubert: The Making of a Master* (New York: Atheneum, 1967), 17.

44. Ibid., 142.

45. Charles Baudelaire, *Les Fleurs du Mal* (Paris: Éditions Garnier Frères, 1959), 6.

46. Translated by the author from Jacques Crepet, ed., *Les Plus Belles Pages de Charles Baudelaire* (Paris: Éditions Messen, 1950), 309.

47. Charles Darwin, "The Origin of the Species," in *The Heritage of Western Civilization*, vol. 2. 7th ed. (Englewood Cliffs, N. J.: 1991), 228-29.

48. Charles Darwin, "The Descent of Man," in *Heritage of Western Civilization*; vol. 2, 7th ed., 233.

49. Ibid.

50. Mandlebaum, 230.

51. Darwin, *Descent*, 233.

52. Notes taken by the author at a lecture by Professor Daniel Todes at University of California, Berkeley; Spring 1982.

53. Ibid.

54. Manuel, 289-90.

55. Ibid., 292.

56. Leszek Kolakowski, *The Alienation of Reason* (Garden City, N. Y.: Anchor Books, 1969), 62.

57. Ibid., 72-75.

58. Tracy B. Strong, *Friedrich Nietzsche and the Politics of Transfiguration* (Berkeley: University of California Press), 77.

59. Ibid., 74.

60. Friedrich Nietzsche, "The Gay Science," in *The Portable Nietzsche* (New York: The Viking Press. 1954), 95-96.

61. Martin Heidegger, "The Word of Nietzsche: God is Dead," in *The Question Concerning Technology and Other Essays* (New York: Harper Torchbacks, 1977), 58-59.

62. Eric Heller, *The Disinherited Mind* (New York: Farrar, Straus, Giroux, 1957), 93.

63. L. J. Rather, *The Dream of Self-Destruction* (Baton Rouge: LSU Press, 1979), 14.

64. Nietzsche, "The Gay Science," 95-96.

65. Heidegger, 62.

66. Ibid., 106.

67. Mandelbaum, 339, 342.

68. Ibid., 343-44.

69. Ibid., 347.

70. Strong, 56-77.

71. Mandelbaum, 341, 347.

72. Nietzsche, *The Birth of Tragedy and the Case of Wagner* (New York: Vintage Books), 72, 99.

73. Ibid., 86, 94-5.

74. Strong, 244.

75. Nietzsche, *The Joyful Wisdom* (New York: Fredrick Ungar Publishing Co., 1960), 294. Please note that *"The Joyful Wisdom"* is given as the title of *Die Froeliche Wissenschaft* in this edition, rather than *"The Gay Science,"* employed in another edition and quoted above.

76. Mandelbaum, 340, 343.

77. Nietzsche, "Thus Spake Zarathustra," in *The Portable Nietzsche,* 428.

78. Mandelbaum, 347.

79. Ibid., 343.

80. Nietzsche, *The Will to Power*, ed. Walter Kaufman (New York: Vintage Books, 1968), 44.

81. Strong, 69.

82. Nietzsche, *The Will to Power*, 345.

83. R. J. Hollingsdale, *Nietzsche: The Man and His Philosophy* (Baton Rouge: The LSU Press, 1965), 88-91.

84. Heller, 87.

85. Hollingsdale, 312.

86. Heller, 125.

87. Strong, 240-41.

88. Heller, 46-47.

89. Ibid., 76.

90. From notes taken by the author at a lecture by Professor Martin Jay at U.C. Berkeley, Winter 1982.

91. Nietzsche, *The Will to Power*, 51, 460.

92. Ibid., 41-42, 52.

93. Ibid., 75.

94. Ibid., 42, 58-59, 62.

95. Nietzsche, "Thus Spake Zarathustra," 398.

96. Strong, 75.

97. Walter Kaufman, *Nietzsche: Philosopher, Psychologist, Antichrist* (Princeton: Princeton University Press, 1950), 36, 40-41.

98. Ortega y Gasset, *Meditación de la técnica y otros ensayos sobre ciencia y filosofía* (Madrid: Revista de Occidente en Alianza Editorial, 1982), 80-82, 87.

99. Jan Sebestik, "The Beginning of Technological Thinking in the Late Eighteenth and Early Nineteenth Centuries." A lecture given at the University of California, Berkeley, 1 Dec. 1981.

100. Iain A. Boal, "A Flow of Monsters," in *Resisting the Verbal Life*, ed. James Brook and Iain A. Boal (San Francisco: City Lights Books, 1995), 4.

101. Lewis Mumford, *Technics and Civilization* (New York: Harcourt, Brace and World, Inc., 1963), 169.

102. Galgan, 304, 315.

103. Ibid., 305.

104. Ortega y Gasset, 53.

105. Noble, 83, 85.

106. Ibid., 83.

107. Langdon Winner, *The Whale and the Reactor: A Search for Limits in the Age of High Technology* (Chicago: University of Chicago Press, 1986), 13.

108. Karl Marx, "The Economic and Philosophical Manuscripts," in Erich Fromm, *Marx's Concept of Man* (New York: Felix Ungar Publishing Co., 1969), 48, 51.

109. Ibid., 98-9.

110. Bluhm, 423-25, 429-30.

111. Marx, "Critique of the Gotha Program," in *The Marx-Engels Reader*, 531.

112. Lewis Feuer, *Marx and the Intellectuals* (Garden City, N. Y.: Anchor Books, 1969), 165-66, 169-71.

113. Engels, "The Origin of the Family, Private Property, and the State," in *Karl Marx and Friedrich Engels: Selected Works in One Volume* (London: Lawrence and Wishart, 1968), 508.

114. Ibid., 511-12, 517.

115. Charles Fourier, *The Utopian Vision of Charles Fourier*, ed. Jonathan Beecher and Richard Bienvenue (Boston: Beacon Press, 1971), 284.

116. Russell Means, "Fighting Words on the Future on the Earth," in *Questioning Technology* (Philadelphia: New Society Publishers, 1991), 73

117. Ibid., 71, 75, 77.

118. Heller, 99.

119. Ibid., 106.

120. Ibid., 107.

121. Nietzsche, *The Will to Power*, 43.

122. Nietzsche, *The Joyful Wisdom*, 308.

123. Nietzsche, *The Will to Power*, 464.

124. Nietzsche, "Thus Spake Zarathustra," 414.

125. Nietzsche, *The Joyful Wisdom*, 339-40.

126. Ibid., 285.

127. Charles Rosen and Henri Zerner, "Enemies of Realism," *New York Review* of *Books*, 4 Mar.1982, 31.

128. Starkie, 282.

129. Rosen and Zerner, "What is, and is not, Realism?" *New York Review of Books*, 18 Feb. 1982, 21.

130. Rosen and Zerner, "Enemies of Realism," 30.

131. Rosen and Zerner, "What is, and is not, Realism?" 24.

132. Baudelaire, *Les Fleurs du Mal*, 36.

133. Habermas, 8-9.

134. Ibid., 9.

135. Starkie, 257-9.

136. Starkie, *Baudelaire* (London: Victor Gallanz Ltd., 1957), 258.

137. Ibid., 259.

138. Ibid., 267.

6
THE TWENTIETH CENTURY

The New, German Theory of Autonomous Technology

THE CONSERVATIVE RIGHT'S MISGIVINGS ABOUT TECHNOLOGY IN THE WEIMAR PERIOD

The reactionary current in pre–World War II conservative German thought, in which a new critical but ambiguous theory of technology appeared, was largely a response to the negative experience of the Weimar Republic. It lasted from 1919 to 1933, when Adolph Hitler and the National Socialists came to power. During the Weimar period, Germany had to accept the humiliation of the Treaty of Versailles, France occupied the German territory of the Ruhr to extract war reparations, the catastrophic inflation of 1923 occurred, and the onset of the Great Depression at the end of the decade brought on high levels of un-employment. The bickering of political parties and the inability of the Weimar government to deal effectively with the nation's problems was another factor that contributed to political extremism on both sides of the political spectrum. The result was, as Jeremy Herf put it, "a republic without republicans," in which the genesis of an antidemocratic outlook and the emergence of romantic anti-capitalism occurred. Both of these attitudes were characteristic of the German conservative revolutionaries of the right at that time.[1]

A number of these conservative thinkers had misgivings about technology as well. For example, Ludwig Klages wrote: "The machine can destroy life but never create it." He warned of a growing domination of mind over soul, and he considered science and technology to be the basis of new myths that promoted the seductive view that they were synonymous with natural phenomena.[2] Ernst Nickish, who had fought in World War I, wrote of the "rape of nature" by technology and said: "Technology murders life by striking down, step by step, the limits established by nature . . ."[3] Despite the misgivings of these thinkers, however, the dominant tendency on the right, represented by the reactionary modernist philosophers, was to reify technology and to glorify and accept it, regardless of the dangers or disadvantages it brought.

THE REACTIONARY MODERNISTS' NEW PHILOSOPHY OF TECHNOLOGY

The fateful transfer of technology from **Zivilisation** *to* **Kultur.** Leading German thinkers of the nineteenth century like Hegel and Nietzsche had rejected the core ideas and values of *Zivilisation*, which they had associated with France and the Enlightenment. These ideas included a universalizing theory of human nature and the promotion of abstract principles like democracy and equality at the expense of concrete institutions and traditions which, as Hegel had recognized, were the historical product of the development of a particular society or people. According to Herf, this German preference for the particular, *Kultur*, derived from the fact that until the latter stage of the nineteenth century, Germany was politically fragmented into separate principalities where the people were rooted in their native soil and own ways of thinking and acting. The French, on the other hand, had lived in a nation-state for centuries and had had a monarchy that created "styles."[4]

Nietzsche, as we noted previously, associated science and technology with what he considered to be the "reactive" values of *Zivilisation* and did not believe, therefore, they could provide a solution to the cultural crisis facing Europeans. In the twentieth century, however, reactionary modernist philosophers and the National Socialist ideologues, despite their affinity with much of Nietzsche's thought, took the fatal step of transferring technology to the ideological category of *Kultur*, which they saw as the valid alternative to the corrupted values of *Zivilisation*.

Both reactionary modernists and Nazi ideologues combined the old—land, blood, history, and tradition (i.e., *Kultur*)—with the new—modern technology. They were not the first, however, as we have seen, to make a synthesis of chronologically disparate elements. Medieval Christian theologians had associated technological advance with recovery of prelapsarian human qualities. Rousseau had been modern in promoting democracy and submitting a blueprint for changing society for the better in his *Social Contract*, but he had also espoused pastoral values and rejected the idea of progress in his *Discourses*. Similarly, Marx and Engels's communist utopia was to be based on a combination of modern industrial technology and social forms borrowed from the earliest, barbarian period in history.

Sombart's blame of capitalism and the Jews for the perversion of technology. Werner Sombart, who published two works in 1911, *The Jews and Economic Life* and *Technology and Culture*, commented on the negative aspects of modern technology, although he did not think such characteristics were inherent in it. He believed, rather, that in defiance of its rightful affiliation with *Kultur*,[5] technology was being misused in a capitalist economic system and subverted by the Jews, who represented a calculating, bourgeois spirit. Pre-capitalist technology, he said, was based on custom and tradition; but in the Jewish-dominated capitalist economy, it involved the application of scientific theory to specific problems. Jewish spirit, he added, was opposed to primordial German values, which included labor and technological creativity. The result was that commerce had gained the upper hand over technology. The solution to this problem, in his view, was to give primacy to technology but incorporate it into a Christian, Germanic, and anticapitalist world of use, not exchange, value.[6] In other words, Germans should combine the old, traditional values, with the new, modern technology.

Sombart identified World War I as a turning point that revealed to him the use of technology outside of exchange relations and showed clearly the positive value it had lost in the commercial world. In the war's aftermath, he described technology as a "true child of the revolutionary, European, Faustian spirit," inseparable from modern science, possessing an "immanent tendency to limitless and almost automatic expansion" of its knowledge, and making emancipation from the "limits of living nature possible."[7] This sounded much like a reifying treatment of technology; but in contrast to the other reactionary modernists, Sombart believed that the nature of the political economy and racial factors determined whether technology appeared in a true or perverted form.

Spengler's conflicted embrace of technology, reified as a thing in itself. Oswald Spengler, the most important of the reactionary modernists, clearly identified some of the most disturbing negative aspects of technology. In his famous work, *The Decline of the West* (1918), he warned of the "devilish nature of the machine" and of the danger of "enslavement" of humans by their own technological creations.[8] Like fellow reactionary modernists Ernst Jünger and Carl Schmitt, but in contrast to Sombart, he reified technology and gave it the ontological status of a thing in itself.[9] For him, in effect, technology had taken on an autonomous character and was bringing about an inversion of the means/end, technology/human relation. He saw a threat that it would ultimately dominate the human will that created it and deny humans control over their own history.[10]

As in the case of Sombart, however, the phenomenon of World War I led Spengler to embrace technology and to see in it the solution to Germany's problems despite his misgivings about it. If the experience of the political failures and economic chaos of the Weimar period gave rise to the anti-democratic and anticapitalist outlook of the reactionary modernist thinkers, his and Sombart's writings showed that World War I was the historical event that proved to the reactionary modernists that technology belonged unequivocally to the authentic, Germanic realm of *Kultur*.

Thus Spengler declared in *Decline* that *Kultur* found its true expression in war, the radical alternative to bourgeois culture. The war, he said, revealed the fundamental truth that it was the "creator of all things." War and production, technological enterprises par excellence, were for him creative activities that had an aesthetic dimension.[11] He declared that the most important work of art to emerge from the trenches, however, was a "new man," anti-bourgeois and, in words that had a Nietzschean ring, a "beast of prey."[12]

Despite this embrace of technology, Spengler did not abandon his concerns about its dangers. He admitted, for example, that the machine "revolted" against Nordic man. The human master of the world, he said, was becoming a slave of the machines he had created. Furthermore, he saw the process of technological development as leading away from its vitalist origins, toward an existence separate from nature, and threatening to undermine the soul.[13] Yet faced with these contradictions, he nonetheless counseled stoicism and heroic surrender to fate and technology,[14] much as Nietzsche had advocated acceptance of the eternal recurrence, with all the joys, sorrows, and pains of existence.

In a later work, *Man and Technics* (1931), Spengler added a Eurocentric and racial dimension to his analysis, stating that only Europe had the cultural resources to develop the machine. The nonwhite part of the world, he stated, had no need to do so and saw technology only as means.[15] Although this comment was not anti-Semitic, it was consistent with Spengler's commitment to German *Kultur* and anticipated the ethnocentrism and racial doctrine that would come to pervade National Socialist ideology.

Spengler's promotion and Ortega's questioning of the engineer's role in resolving the cultural crisis in Germany. To restore Germany to its rightful place in the world, Spengler called for a program of German technological advance carried out by a community of blood, which he equated to socialism. Political liberalism and capitalist devotion to money would be replaced by an authoritarian state in which all class conflicts would be transcended.[16] In *Prussian Virtue and Socialism* (1919), Spengler explicitly transferred socialism from *Zivilisation* to *Kultur* by combining it with German nationalism. He praised the Prussian virtue by which the individual served the whole, and he called for a "dictatorship of organization" to maximize national power.[17] With ideas like these, he prefigured the totalitarian outlines of the National Socialist regime under Adolph Hitler. By reconciling German nationalism, socialism, and technology, Spengler and other reactionary modernist thinkers laid the groundwork for the ideology of their National Socialist successors.

Spengler reserved a special role for engineers in his proposed program of national socialism. The autonomous, "Faustian character" of technology, he said in *Decline*, gave it an irrational and romantic dimension, a "satanic quality" that could only be understood by those who had the most affinity with technology— engineers. They had to take the lead in Germany, he declared, and provide the "guidance and leadership" necessary to integrate technology with a new culture.[18]

His focus on the engineer's role found an echo in the articles that engineers of the reactionary modernist persuasion, like Heinrich Hardenstett, Karl Weihe, and Manfred Schroter, wrote in the 1920s and 1930s for German professional journals. In a similar vein, National Socialist ideologues deliberately made an appeal to German engineers, many of whom gave support to Hitler after he came to power. The irony, however, was that the Frenchman Auguste Comte, whose positivist philosophy of the nineteenth century was a reaffirmation of the En- lightenment values rejected by reactionary modernist thinkers, also had assigned the engineer a crucial role in the creation of a new and better society.

Writing from exile in Argentina at the time, Spain's leading philosopher, Ortega y Gasset, provided a critical response to reactionary modernist thinkers like Spengler and their National Socialist successors who asserted that the en- gineer would play a key role in leading Germany, and Europe, out of the cultural crisis. According to Ortega, who had studied in Germany, the engineer's func- tion was to take charge of "technological projects," which provided humans with enough freedom from nature to have a human as well as a biological life. The question of what to do with this freedom, however, a matter of what Ortega called "vital projects," was one that he said the engineer was not qualified to answer. Vital projects, he declared, were the province of philosophers, poets, artists, mystics, and statesmen. "The engineer," he pointedly concluded, "cannot rule."[19]

Jünger's assertion that the World War I battlefield revealed the beauty and morality of technology. The German writer who played the most important role in romanticizing the battlefront experience of World War I and using it to glorify technology and the machine was Ernst Jünger. His writings at that time included such Nietzschean themes as the praise of will and the need for an exceptional elite with a special mission. He realized, however, that Nietzsche's conception of *Kultur* excluded technology. Thus, he wrote that Nietzsche had no room for the machine "in his Renaissance landscape"; but he added this com- ment: "He taught us that life is not only a struggle for daily existence but . . . for higher and deeper goals. Our task is to apply this doctrine to the machine."[20]

Like Adolph Hitler, Jünger had fought in World War I and participated in the famous battle of Langemarck, in which 145,000 humans died.[21] For him, as for Spengler and Sombart, the war was the event that made the true meaning of technology clear. It showed, said Jünger, that a "misguided romanticism" had characterized the machine as conflicting with *Kultur*.[22] For him, however, the battlefield of World War I revealed the incredible beauty of the technology of destruction. The conclusion he drew from the war, therefore, was that the aestheticized morality of the fin-de-siècle art-for-art's-sake movement, according to which the beautiful was the good, applied also to technology. Since technology had a beauty of its own, he reasoned, it needed no external moral justification. Walter Benjamin commented, appropriately, that Jünger had transformed the thesis of art for art's sake into production for production's sake and destruction for destruction's sake.[23] He should have added to his list, however, its transformation also into technology for technology's sake.

The front generation came to realize, Jünger said, that technology was not only useful, but beautiful—that is, that it was a source of higher satisfaction. Because of the war, he said, the artistic individual was suddenly able to see a totality in technology, the presence of two different orders or goals. The engineer and the socialist, Jünger added, both recognized the validity of this "solution."[24] For them, it was apparent that technology belonged to *Kultur*, not *Zivilisation*.

Jünger's call for surrender to technology, akin to nature and good because it crushed the individual. War was an experience, Jünger declared, where the individual asserted himself in battle but at the same time submitted to forces beyond his control.[25] In a remarkable passage about a battleship "sinking with flags flying," Jünger described its elaborate technical organization and machinery. "This whole apparatus," he wrote, "is sacrificed in seconds for the sake of things which one does not know but rather in which one can only take faith."[26]

This quasi-mystical glorification in Weimar and Nazi Germany of faith in the absurd and the technologically efficient activity of war had a familiar nihilistic ring. Indeed, the ideas of submission to fate and individual self-sacrifice in meaningless but technologically perfected activity done essentially for its own sake and without regard for ultimate purpose or consequences. They reflected the same values as Dostoevsky's Sniveling Hero, who declared he loved the process of the game more than its outcome, and Nietzsche's overman, who was to be strong enough to face up to the pains and sorrows of existence without any hope of redemption.

Jünger's idea of surrender to the overwhelming fatality of an absurd but technologically efficient process was similar to Spengler's exhortation to Germans in *Man and Technics* to surrender to fate and technology, despite the dangers. It was likewise an expression of the nihilistic values to be found at the very core of the National Socialists' political praxis, particularly after the war began. As Arendt later pointed out in her famous analysis of Nazism in power,

Totalitarianism, motion itself was the very essence of the Nazi phenomenon.[27] The Reich, she said, was characterized by "shapelessness," with a duplication of functions, a constant shifting of the locus of state power, a lack of interrelation among various ruling strata, swift and inexplicable shifts in policy, a neglect of material interests, and a preoccupation in occupied areas with the extermination of subject peoples rather than their exploitation.[28] Caught up in such an environment of absurdity and constant flux, the individual had no firm grounding and essentially was lost.

In agreement with Spengler, Jünger recognized that modern technology had a dehumanizing effect. And like other reactionary modernists, he saw this assault on the individual as a positive development that undermined one of the most precious achievements of the Enlightenment.[29] Locke's political theory, for example, had emphasized the rights of the individual, as had the French Revolution's *Declaration of the Rights of Man and Citizen*. A key revelation of the Battle of Langemarck, said Jünger, was that it showed that the "individual, representing all that is weak and doomed," was crushed by the "steel laws" of the mechanical age.[30]

The war also convinced Jünger that scientific and technological progress led not to peace, but to total mobilization. In Jünger's view, total mobilization was the reflection of a global trend toward state-directed activities in which the individual would be sacrificed to the demands of authoritarian planning.[31] Thus he, like Spengler, provided a rationale for the emergence of the totalitarian Nazi state.

In *Storm of Steel* (1920), Jünger likened war to a conflict of natural forces. An artillery barrage was for him a "storm of steel," and a bomber was a "vulture."[32] Similarly, he described technology itself as an expression of the darker side of human nature. Technology, he said, was an instrument of human will, the expression of a "mysterious and compelling claim."[33] The "hot will of blood" and the power of technology were, in his view, joined in battle.[34] The war experience showed, he declared, there could be no "false return to nature" (i.e., to the Romantics' benevolent and idealized conception of it), because nature and technology were not opposed.[35]

NATIONAL SOCIALIST ADOPTION OF THE REACTIONARY MODERNIST PROGRAM, INCLUDING ITS GLORIFICATION OF TECHNOLOGY

When Hitler came to power, he tried to carry out the cultural program the reactionary modernists advocated.[36] He was, in Herf's view, the most important practitioner of the reactionary modernist tradition and the one who "started the war which was to unify technology and the soul."[37] Not surprisingly, Hitler appointed Albert Speer, an architectural engineer whose expertise combined aesthetics and technology, to be his chief advisor. Speer orchestrated the elaborate and aesthetically striking mise-en-scènes for Hitler's mass rallies and, in his own words, "designed the buildings and produced the weapons which served his [Hitler's] ends."[38]

On both a practical and an ideological level, the National Socialists emphasized the value and importance of technology. In 1936, Hitler announced a four-year plan of economic development. The Nazi economic policy focused on the goal of reducing German dependence on the rest of the world by means of technological innovation.[39] Fritz Nonnenbrach made use of a typical Nazi vocabulary when he described the development of synthetic fuels by the German chemical industry as the expression of a "will to freedom" that would liberate the nation from reliance on foreign raw materials.[40] Fritz Todt, a party member with close ties to engineers, described the creation of a national highway system (the *Autobahnen*) as a way to provide a unity that would override the chaos of Weimar.[41] Goebbels, the minister of propaganda in the Third Reich, wrote in 1939 that the development of the Volkswagen showed that under Hitler modern technology was accessible to the masses.[42]

The Nazis reified technology as a thing in itself by linking it to nature and giving it an autonomous will. For them, however, as for the reactionary modernists, will was not the rational factor that Rousseau and Kant considered it to be. It was, rather, an irrational element, as Schopenhauer and Nietzsche had defined it.

The inclusion of the racial factor in National Socialist ideology was an important example of this emphasis on natural factors as part of true culture. Hitler linked his anti-Semitic theories to so-called biological laws, and the Nazi treatment of the technology question had a racial dimension as well. For example, Hitler wrote in *Mein Kampf* that Aryan culture equaled Greek spirit joined with German technology.[43] Ferdinand Fried, a Nazi ideologue, wrote in *Deutsche Technik* that Germany's racial soul was in tune with technological advances, although technology had been "raped" by the Jewish *Ungeist*.[44]

Following the lead of reactionary modernist thinkers like Jünger and their sympathizers in the engineering profession, National Socialist ideologues aestheticized technology. This was an important point for them, not only because it was a way of showing that technology belonged to *Kultur*, but also because it was consistent with their contention that since technology was beautiful, it was undeniably good. Pro-Nazi articles published in the engineering journal *Deutsche Technik* pointedly referred to "Goethe the technologist" and traced technology back to Leonardo da Vinci, whom they cited as a model of the gifted engineer-artist type.[45]

After the war's conclusion, Hitler's top aide, Albert Speer, was convicted by an international tribunal of crimes against humanity and sentenced to a life term in Spandau prison. Although he was an architectural engineer, he pointedly, and belatedly, expressed misgivings about technology in his memoir, *Inside the Third Reich* (1970, English edition). "Dazzled about the possibilities of technology," he wrote, "I devoted crucial years of my life to serving it. But in the end, my feelings about it are highly skeptical."[46] He pointed out that modern technology made it possible for a state apparatus to be subjected to direction by a single will and said that Nazi Germany was the "first dictatorship . . . which employed to perfection the instruments of technology to dominate its own people."[47] Furthermore, he issued a warning: "The more technological the world becomes, the greater is the danger . . . There is nothing to stop technology and science from completing its work of destroying man which it began in this war . . ."[48] One has only to remember the megadeaths resulting from the atom bomb attacks on Hiroshima and Nagasaki in 1945, to cite a well-known example, to appreciate that the experience of World War II provided ample ground for such a warning.

THE SPECIAL CASE OF HEIDEGGER VIS-À-VIS TECHNOLOGY AND NATIONAL SOCIALISM

Heidegger's perception of the inadequate response of National Socialism to threat posed by technology. In 1933, Martin Heidegger, who was Germany's most prominent philosopher and a critic of technology of the conservative right, was appointed rector of the University of Freiburg several weeks after Hitler had seized dictatorial powers. Heidegger shared some of the conservative values of the reactionary modernists but not their enthusiasm for technology and their belief that it belonged to the realm of *Kultur*.

For Heidegger, the European civilization in which technology developed had been in a process of decay ever since the Greeks had initiated, by his assessment, an active, dominating stance vis-à-vis nature.[49] By this he meant that they had started a philosophical tradition involving the "anthropologizing of Being," according to which human reason assigned to everything in the world its place, purpose, and meaning instead of remaining "open to Being" as it was revealed in the flux of time.

On the level of the actual physical domination of nature, his assignment of blame to the Greeks immediately brings to mind the Prometheus myth; but he ignored the fact that Greek *techne* was one of limits. And in the case of natural philosophy, he ignored the fact that Apollonian science was one of explanation and could not be practically applied for the domination of nature. Similarly, because of his attachment to the "anthropologization" thesis, he also ignored the fact that the ideological roots of the glorification of technology were in Christian theology, not classical Greek philosophy.

In any case, when the National Socialists came to power, Heidegger hoped they would provoke a crisis in which the "forgetting of Being," abetted by technological advancement, would be overcome.[50] In his inaugural address at the university, Heidegger inconsistently, but not without opportunism, chose to name Prometheus, the one who stole fire from the gods and taught the secret of technology to humans, as "the first philosopher." By his actions, Heidegger said, Prometheus handed himself over to an "overpowering fate" and showed that he correctly understood that theory was praxis—although not the decadent praxis of the Marxist variety. At the decisive moment in history before Germany, Heidegger said, there would be a Promethean theory (i.e., National Socialism) that would come to grips with fate and help to unify the German people.[51]

Within a year, however, Heidegger had second thoughts about the Nazi regime and stepped down from his rectorship. Both the U.S. and the USSR., he believed at the time, were fostering a "wild and endless race of unleashed technology," for which there had to be an alternative. But the Nazi regime, rather than halting technological development, as he had hoped, soon appeared to him to be perverted by technology.[52] After his resignation, Heidegger was placed under surveillance by the Nazi government. He retained a teaching post at the University until the war's end, however, when he lost it on the order of the Allies' postwar Denazification Commission.[53]

In the mid-1930s, Heidegger published several essays in which he elaborated on his conception of technology and its relation to Nazism. Rather than concerning himself with the social context within which technological innovation took place or the political responses to such innovation, however, he reified technology and preferred to speculate, like the reactionary modernists and thinkers of the conservative right, on its essence or spirit.[54] In effect, he saw technology as an autonomous, self-steering force that threatened the individual.[55]

Heidegger's definitive, postwar statement on technology. Several years after the German defeat, in the winter of 1950–51, Heidegger was allowed to resume lecturing. The following year, he formally retired from the university.[56] In this early postwar period, however, he published a number of books and papers on philosophical questions. These included *A Letter on Humanism* (1946),[57] his polemic with the leading French philosopher of the postwar period, Jean-Paul Sartre, and his well-known essay, *The Question Concerning Technology* (1953), where he focused again on the topic he had first discussed in *Time and Being* (1927). Despite the effects of the political turmoil of the Weimar, Nazi, and early postwar periods, Heidegger's thought endured and had an undeniable influence, for better or worse, on contemporary Western philosophy.

In *The Question*, Heidegger made a definitive reply to the question of technology's philosophical significance and essence, removed a generation in time from the reactionary modernists and National Socialists who had addressed the same topic but had given answers that he rejected. He did not retreat from his original view that modern technology incorporated an inherent will to power; and he defined technology's essence, in his peculiar and sometimes obscure philosophical language, as "enframing." It was a way of ordering—and revealing—the world, he said, so that everything was made into a "standing reserve" that no longer stood against humans as objects.[58] To put this in less specialized terminology, we can say that for Heidegger, the essence of technology was that it rendered things external to humans objects of their manipulation and control. Technology, in other words, did not leave anything free to actualize, or reveal, its own true being.

Heidegger warned in *The Question* of a twofold danger because of the exaggerated importance of technology and its advanced stage of development. First, he said, at the very point humans had exalted themselves into the posture of lords of the earth, they might find that they, themselves, were being taken as a "standing reserve." That is, the fact that humans, themselves, were increasingly the objects of technological control could prevent them from freely revealing their own beings as they existed in the flow of time. And second, the "enframing" effect of their technology threatened to deny them the possibility of entering into a more original mode of revealing being in general and "experiencing the call" of a more primal truth about it. The essence of technology, therefore, was ambiguous: It let humans endure, but at the same time it blocked the kind of authentic revealing which related to truth and to being.[59]

Heidegger's suggestion that art might provide an answer to the threat of technology. To suggest a possible solution to the dilemma posed by technology, Heidegger pointed out that for the ancient Greeks, the *poesis* of the fine arts was also *techne*. Art, however, had always had a privileged role, he added, because it was a "more primarily granted" form of revealing. For that reason, he asserted, it could foster the growth of a "saving power" in relation to truth and rescue humans for the role they were destined to play in the safeguarding of Being. A decisive confrontation with technology, therefore, would have to occur in this realm, akin to its essence yet different.[60]

To rephrase this idea in plainer language, we can say that art relies on a creative and imaginative faculty—that is, sensuous reason—to discover and to enrich the world. Instead of the enframing associated with technology—that is, the taking of everything as objects for domination—art opens up new ways for things to "be" and thus contributes to the revelation of their essential truth. For Heidegger, therefore, art offered a possible response to the philosophical crisis that afflicted a civilization perverted by too much technology.

Unfortunately, art in the twentieth century, rooted in Kant's proclamation of its philosophical autonomy at the end of the eighteenth century and influenced by the theoretical innovations of the art-for-art's-sake movement and the Realist school of the nineteenth century, ultimately abdicated this "saving" role and, essentially, merely reproduced the philosophical void that prevailed in the culture at large.

Pablo Picasso (1883–1973), perhaps the greatest plastic artist of the twentieth century, provides an instructive example of the inability of twentieth century art, despite its unquestioned achievements, to satisfy Heidegger's ultimate expectations. Picasso symbolized the autonomy of the modern artist in his experimentation with different types of subject matter and styles, including his blue period, rose period, Cubist phase, collages, junkyard sculptures, Mediterranean-motif ceramics, tauromachy and minotaur drawings, and inflated and contorted representations of women and children. Indeed, in some of the latter, he distorted his subject matter to a point of utter nonrecognition, and even negation, thus showing the ultimate subordination of all content to the will of the modern artist. Furthermore, in spite of a certain commitment to humanism and his Mediterranean roots, he personified the idea that the process of change in styles was the true subject of modern art. Picasso admitted that he was more interested in questions than answers, and his own never-ending journey across a range of genial stylistic experiments and subjects ultimately confirmed the absence of any durable philosophical truths in his century.

Post–World War II Theory:
The Technological Society and Its Critique

In the first half of the twentieth century, Heidegger and a number of other intellectuals of the conservative right in Germany had advanced the idea that technology posed a threat to human well-being. This was a definite departure from the traditional ideological view, rooted in the Christian theology of recovery and Judeo-Christian millenarianism, that technological progress would be part of the realization of human destiny. Heidegger, however, as we have seen, reified technology and situated his critique of it within the narrowing context of his ontology. That is, he included none of the analysis of the social, political, and instinctual aspects of modern technology that characterized the critical writings of thinkers like Jacques Ellul and Herbert Marcuse after World War II. The reactionary modernist philosophers and Nazi ideologues who succeeded the technology critics of the conservative right in Germany, despite their appreciation for technology's negative effects on both humans and nature, ultimately accepted it, and even glorified it. Essentially, they saw modern technology as an expression of their nihilistic values and as the practical means to attain their nationalistic goals.

The opening of social thought and philosophy to a critique of technology in the twentieth century, in any case, was a logical consequence of its mounting power and sophistication, as well as its increasingly visible harmful effects, including environmental destruction, dehumanization of workers and citizens, and awesome destructive power as revealed in war. It was also a logical sequel to Hegel's philosophical examination of the limitations of scientific reason (*Verstand*) and Nietzsche's assessment of science as a false and reactive will to power in the nineteenth century. After their philosophical critique of science, it was inevitable that technology, increasingly the practical application of theoretical scientific knowledge, would likewise be subjected to critical scrutiny in terms of philosophical implications and social and cultural consequences.

With the publication of *The Failure of Technology* more than two decades before Speer's denunciation of technology in his *Inside the Third Reich*, Friedrich Jünger, the youngest brother of reactionary modernist Ernst Jünger, made an important contribution to a postwar critical trend.[61] His work was a reprise of the critique by the antitechnologists of the Weimar conservative right and a fitting epitaph to the demise of technology-glorifying Nazi Germany. One of the least original aspects of this retrospective analysis, however, was that despite his rejection of modern technology, Jünger reified it, as the reactionary modernists and Nazis had done, treating it as a thing in itself.

Also in the mid-1940s, Frankfurt School thinkers Max Horkheimer and Theodor Adorno, exiled in the United States during the war, prepared the ground for a forward-looking critique of the postwar sociocultural order with their famous work, *Dialectic of Enlightenment* (1947). They combined in their so-called "critical theory" elements of Marx's analysis of political economy, Freud's theory of the instincts, and a demonstration of how the prevailing positivist thought, based on scientific methods and standards of truth, spared the existing social and cultural order any effective ideological challenge.[62] In the 1960s, Marcuse used these same philosophical tools to make an incisive analysis (*Eros and Civilization*, *One Dimensional Man*) of the instinctual and philo-sophical dimensions of what he called "advanced industrial society."

The most comprehensive of the new, postwar technology critiques, however, was that of the French professor of law and sociology, Jacques Ellul. He published his noted work, *The Technological Society*, in 1955. Although this analysis applied primarily to the technology of the late industrial age and the social transformations it had brought about in the first half of the twentieth century, particularly from the 1930s to the early 1950s, it turned out to be the most powerful and thoroughgoing critique of technology published in the second half of the twentieth century. Essentially, it was a phenomenological analysis of the role, operation, governing principles, and effects of modern technology in the new social order it had come to define, what Ellul called the *technological society*. This name reflected his belief that the impact of modern technology was so pervasive that it had to be treated as a totalizing sociological phenomenon.

When Ellul initially published his work in 1955, the essential features of the technological society could be found in industrial nations in Eastern and Western Europe and North America. In these societies, not only were goals, norms, and procedures determined by technological criteria, but technology itself, instead of being oriented by higher values, as had been the case for Greek *techne*, was essentially subject only to its own narrowly defined standards and purposes.

As Postman accurately remarked about the kind of technological social order Ellul was describing, "[T]he traditional was still there."[63] Yet Ellul was not unaware at the time he wrote his *Technological Society* that a process of subversion of traditional culture by technological goals and values was underway. This transformation would not be completed until the end of the second Christian millennium, by which time the spread of information technology and the impact of global communications media had had their effect. At that point, postmodern technoculture,[64] or what Postman referred to as *Technopoly*,[65] had become a reality in some of the most affluent and technologically advanced societies of the West.

Although many of the values and institutions of traditional society and culture remain in the poorer nations of the world today in Asia, Africa, the Middle East, and Latin America, often euphemistically referred to as "emerging economies" by the dominant actors in the global economic arena, important features of the technological society and even signs of technoculture can be found in them. Westernized technocrat elites frequently rule in these nations; and modern technology—from surface-to-air missiles, laptop computers, and antibiotics to genetically-engineered crops, satellite television, and modern technological norms and methods of administration—has made significant inroads, even if indigenous social structures have not eroded to the point where one could say that a full-blown technological society exists. And thanks to global communications media and information technology, the cultural symbols and virtual realities of technoculture are likewise invading these societies, although limited personal access to these technologies has thus far restricted the scope of their impact.

ELLUL'S PHENOMENOLOGICAL ANALYSIS OF TECHNOLOGY IN THE TECHNOLOGICAL SOCIAL ORDER

The originality of Ellul's analysis. Ellul presented in his *Technological Society* a description of the characteristics of the phenomenon of modern technology, which he dated from the Industrial Revolution and described as qualitatively different from all technology which had preceded it. He stated that modern technology had become the very infrastructure of civilization, representing a "multiplication of means without limit," constantly evolving with disconcerting rapidity. Affecting every sphere of human activity, he said, it had created a global unity of civilization, confronting humans with an ever-increasing array of technology-generated problems.[66] This was a crucial point of difference from Marx and Engels, who had essentially portrayed the advance of technology as a secularized theodicy of guaranteed benefit.

Ellul realized, like Marx, Heidegger, and the reactionary modernists before him, that modern technology had become much more than neutral means in the Aristotelian sense. Furthermore, although a reading of his *Technological Society* reveals that Marx had a great influence on the development of his ideas about technology, there was an important difference in their conclusions. Whereas Marx asserted that the level of technological development determined the level of everything else in the social system, Ellul found that modern technology had attained such importance that *its own* values, methods, and goals had permeated every aspect of society. This was why, he said, a qualitatively different type of society had come into being, what he called a technological society. The formulation of this idea as the basis for his analysis was its most original feature, and it gave the theory of technology a new, critical dimension.

The Augustinian perspective behind Ellul's secular phenomenological analysis. Although Ellul rejected the Christian view that technological advance and the realization of human destiny were linked, it must be noted that there were Christian theological beliefs underlying his analysis of technology as a social phenomenon. That is, it was precisely because of his Augustinian Christian perspective that he associated technology with the human struggle for survival in a degraded nature, rather than with recovery after the Fall. In other words, he shared with Augustine the idea that technology was only of "some use" for humans. Furthermore, quoting Genesis to illustrate that God finished the Creation before the Fall, he pointedly rejected the view that humans had a "demiurgic" function of assisting God in "finishing" it.[67] Despite the lack of a direct reference to these points in *Technological Society*, he did declare that as technology advanced (contrary to the theory of the millennium and Marx and Engels's utopian promise), the facts demonstrated that humans were experiencing a progressive loss of freedom.[68] For him, as he indicated elsewhere, the phenomenon of technology without limits clearly belonged to the realm of necessity, not freedom. Its characteristics were domination and efficiency, he said; and it was a betrayal of Christian love, another example of the sinful abuse of freedom by humans that had started with the Fall.[69]

The analysis in *Technological Society*, however, has been recognized as consistently secular on its face and based on what Ellul considered to be a factual examination of modern technology as a social phenomenon, without regard for underlying causes.[70] Its power, scope, and rigor is so compelling, in any case, that it appealed, and continues to appeal, to many individuals throughout the world who do not share, for either religious or philosophical reasons, Ellul's theological views.

The autonomy of modern technology in practice. Ellul did not reify technology and assert that it has a willful soul or essence, as Heidegger, reactionary modernist philosophers, and Nazi ideologues had done. His analysis of modern technology in practice led him rather to argue that modern technology follows its own internal dynamic and does not represent neutral means amenable to orientation by external values. "External necessities no longer determine technique," he wrote. "Technique's own internal necessities are determinative."[71]

These internal necessities were for Ellul an efficiency calculus, or what he called the logic of "the one best way."[72] It requires, he said, that for any specific endeavor, a technical specialist be called in to find the "one best way," in the absolute sense, as determined by numerical calculation.[73] That is, the operational goal is always to maximize the predetermined result with the least amount of input. This means that quantitative efficiency standards replace qualitative criteria. Considerations that are social, ecological, moral, spiritual, aesthetic, or traditional, therefore, are relegated to a secondary status.[74]

The moral ambiguity of modern technology in its effects. Because modern technology frequently has unforeseeable and undesirable side effects along with intended beneficial results, Ellul concluded that it is morally ambiguous in practice—that is, both good and bad at the same time. In other words, it creates new problems in the process of solving existing ones. Similarly, he pointed out that the disadvantages of a given technology—the creation by power-generating nuclear reactors of dangerous radioactive wastes that must be safely stored for thousands of years, for example—could not be eliminated by changing the nature of the political economy, such as replacing capitalism with socialism, or the contrary.[75] This observation was another example of his rejection of the traditional Aristotelian view that the harm or benefit of a technology depends solely on the choice of use. For these and other reasons, Ellul concluded that it was no longer possible to retain only the good and reject the bad in the case of any given technology.

The inability of social institutions, traditions, democratic processes, and peaceful intentions to determine the nature of modern technology. The belief that a specific technology will be subordinated to the traditional practices and values of the society in which it is used, said Ellul, is contrary to reality. In the first place, he asserted, modern technology penetrates and subverts traditional societies everywhere on the planet, working to incorporate them into the new, universal (now we would call it global) technological society. And furthermore, nothing is sacred anywhere anymore, except technology itself, so existing traditions and social institutions are subject to elimination by quantitative technological standards and efficient practices and are totally expendable.[76]

Modern technology also resists orientation by external values, according to Ellul, because it is no longer merely a part of society but has penetrated and affects every aspect of life and culture.[77] There are no longer, he said, any safe or independent zones of existence. This observation, as well as that noted in the previous paragraph, are examples of Ellul's understanding in the 1950s that technology was working a fundamental cultural as well as sociological transformation, although the former had not been completed at that time.

He also argued that modern technology exhibits in practice an antidemocratic tendency because the criteria for action and methods are no longer determined by free discussion and critique by ordinary citizens but are dictated by technical considerations determined by experts.[78] Furthermore, he asserted, authoritarian state intervention is ultimately needed, regardless of the nature of the political system, to conform everyone and everything to these predetermined technical norms and objectives.[79] In other words, for Ellul, technology has reached a point where it operates independently of the basic values of the political culture of a society instead of being controlled by them. This fact was another confirmation for him that in the modern context, technology can no longer be considered neutral means, subject to orientation by external values.

Modern technology, Ellul further argued, tends to favor war for a number of reasons. Not only does every technology have some kind of military application, he asserted, but modern technology obliterates the difference between offensive and defensive weapons and makes it easier to kill.[80]. (The rationale of "deterrence" for the deployment of nuclear-armed missiles by a superpower during the Cold War, for example, could justifiably have been interpreted by an adversary as a smokescreen for the development of offensive capability.) Ellul thus re-iterated his basic thesis that modern technology is morally autonomous, not because it has a will or a soul of its own, but because it is a phenomenon that consistently exhibits certain forms, follows a specific logic of operation, and produces certain types of effects.

The autonomy of the process of technological innovation. Even the process of technological development, according to Ellul, has essentially become independent of moral direction. To illustrate this point, he quoted the remark by Jacques Soustelle, a post–World War II French intellectual and member of the French political class, that since the atomic bomb was technically possible, it was "necessary." This is the way it is, said Ellul, for all technological development.[81] Potentiality, in effect, dictates actuality. Furthermore, he argued that technological development then implies use. That is, once large amounts of time and money have been expended to produce a new technology, there is an overriding incentive to apply it as rapidly as possible.[82] This rule applies, Ellul added, "without distinction of good or evil," and regardless of whether there is any real objective need for the new technology.[83]

In the same vein, Ellul pointed out that each new discovery makes the next one possible, and that seemingly unrelated discoveries link up with each other to produce new breakthroughs. In such instances, the eventual result is neither predictable nor subject to control. This was another reason why he concluded that the overall process of technological progress has, in effect, no pre-determined goal or finality.[84]

Inversion of the means/end, technology/humanity relation. Summarizing the nature of the interrelation of modern technology in practice with the human world in which it exists, Ellul emphasized the lack of reciprocity that prevails. Although technology is a phenomenon exhibiting freedom and autonomy with respect to the external world, he said, the same cannot be said for the relation of humans to technology. This is the same inversion of the means/end, technology/ human dichotomy that Spengler and Heidegger identified, as we have previously noted. Modern technology in practice is ultimately concerned with only one thing, in Ellul's view: efficient control; yet human beings have to adapt or be adapted to each technological innovation, regardless of their real needs or preferences.[85]

LEGITIMIZATION OF THE TECHNOLOGICAL SOCIAL ORDER

The role of analytical philosophy in eliminating critique. In 1964, Marcuse's *One Dimensional Man* appeared. Published a decade after Ellul's work and purporting to be an analysis of "advanced industrial society," it dealt essentially with the same social order that Ellul had described in his *Technological Society*. Being a philosopher, however, Marcuse placed more emphasis than had Ellul, a sociologist, on the transformation of culture itself by modern technology. In *One Dimensional Man*, he emphasized that there was a new kind of culture to go along with the new, technocentric social order. Without actually resorting to a term like *technoculture*, which appeared in philosophical and sociological discourse only after the information technology revolution at the end of the twentieth century, he wrote: "When technics becomes the universal form of material production, it circumscribes an entire culture; it projects a historical totality—a world."[86]

One of the most egregious mechanisms of this culture was, according to Marcuse, that it effectively excluded critical or conflicting ideas, allowing the established order to become the sole standard and point of reference. In such a one-dimensional cultural context, the ideal was effectively absorbed by the existing situation, the real. Marcuse's explanation of the philosophical developments that contributed to this one-dimensional situation focused on the post-Nietzschean school of analytical philosophy. Its representatives had carved out an influential niche in America and elsewhere in the Anglo-Saxon world, the technologically advanced cultural milieu to which Marcuse gravitated with the rise of Nazism in Germany and the onset of World War II. These analytical philosophers included Ludwig Wittgenstein, Gilbert Ryle, J. L. Austin, and G. E. Moore, all successors to Nietzsche's rejection of metaphysics and history and his conclusion that language was the central problem for philosophers because it created concepts and illusions that humans mistakenly took for reality.

In their view, Marcuse explained, philosophy had to assume a "therapeutic" role. That is, its task was to purge thought of all metaphysics and transcendent concepts, rather than to challenge critically the prevailing universe of discourse and behavior.[87] Accordingly, in what one might consider a trivialization of philosophy and an abdication of moral responsibility,[88] the analytical philosophers concentrated on language and the clarification of linguistic statements. Austin, for example, wrote a wordy text about what it meant to "taste" something; and Wittgenstein applied himself seriously to analysis of the statement, "My broom is in the corner."[89]

Outside of reserving such prosaic tasks for philosophy, the analytical philosophers left it up to scientific inquiry, and the questions that it could answer with its limited objectives and methods, to provide a model for certainty and guarantee the progress of factual knowledge about the phenomenal world. This was essentially a reversion, according to Marcuse, to the legacy of St. Simon, whose positivistic philosophy first began to speak of the world in instrumental—that is, essentially technological—terms in the nineteenth century.[90]

Wittgenstein's declaration that philosophy "leaves everything as it is" and Moore's quoting of Bishop Butler's formula, "Everything is what it is, and not another thing,"[91] were perfect examples for Marcuse of the essential positivism underlying the approach of the analytical philosophers. He considered it one-dimensional, not only because they rejected the traditional negating role of critical thought, based on ideal concepts, but also because they selected the common usage of words, rather than a philosophical metalanguage, as the primary frame of reference.[92]

Legitimization by the empirical evidence of impressive technological achievements. Because of their drastic revision of the character and boundaries of philosophical discourse, only empirical facts were recognized by analytical philosophers as a means to prove the correctness of cognitive thought.[93] Such facts were to be accepted, however, without seeking the deeper meaning that relating them to each other or to a broader context of explanation would bring. Similarly, these analytical philosophers avoided, in Marcuse's view, any truth-revealing critique of existing ideology and social reality. And they failed, likewise, to include a broader perspective of historical development or political economy, within which questions of morality and limits could be properly situated and resolved.

This positivistic emphasis on empirical facts meant that the door was wide open for validation of the advanced industrial social order by technology itself. And the empirical facts created by modern technology and its technological brand of civilization, to say the least, were particularly compelling. They included, according to Marcuse, the "ever-more-effective domination of man and nature ... the ever-more-effective utilization of ... resources,"[94] the enlargement of the "comforts of life," and increase of the "productivity of labor."[95]

Any effective political challenge to the advanced industrial order, Marcuse concluded, was unlikely because the achievements of its technology "veil the particular interests that organize the apparatus." Technology "has become the great vehicle of reification," he added; and the members of the society falsely believe that everything has been determined by "objective qualities and laws."[96]

Similarly, in this new ideological order, there could be no effective challenge to everyday life posed by higher culture, such as literature and the arts, whose traditional role had always been to contradict prevailing social reality. In the new technological order, Marcuse explained, "reality surpasses its culture." In an obvious reference to their technological power, Marcuse declared that humans had reached the point where *they* could "do more than the culture heroes and half-gods."[97]

Thus, although technology occupied a central position in culture of advanced industrial society, it was subject to no external philosophical limits or effective ideological challenge. As Marcuse stated, in theoretical terms, "the transformation of man and nature has no other objective limits than those offered by the brute facticity of matter . . ."[98]

The "magical" appeal of modern technology. The technological social order is not only validated by a continual exhibition of its impressive technological accomplishments and sheltered from ideological challenge by the one-dimensional character of its positivistic philosophical underpinnings, but it is fortified, in Ellul's view, by the quasi-religious function and magical appeal of technology itself. That is, he argued, modern technology fulfills an irrational need of humans to have an omnipotent object of veneration. It is, as he put it, a "marvelous instrument of the power instinct which is always joined to mystery and magic."[99] Humans "cannot live without the sacred," he added. And since "nothing belongs any longer to the realm of the Gods or the supernatural," humans understandably transfer their "sense of the sacred to the very thing which has destroyed its former object, technique itself."[100]

Noble showed in his *Religion of Technology* how Hebrew and Christian theological elements like technological advance as preparation for the millennium or recovery from the Fall, as well as the human-likeness-to-God theme, played an important role historically in promoting the development of technology.[101] What we encounter in Ellul's assessment, however, is something more: the capability of technology to respond to the same psychic needs which religion addresses. Ellul observed that the power of technology is "to the technician an abstract idol which gives him a reason for living and for joy." Humans see it as their salvation, he added, sacrifice themselves to it, and in the end, worship it.[102]

Ellul attributed the worldwide "outburst of frenzy" which greeted the successful launch of *Sputnik*, the first space satellite, by the Soviets in the 1950s to the quasi-religious appeal of technology. He concluded that in the modern world, technology is for humans "the common expression of human power without which they would find themselves poor, alone, naked, and stripped of all pretensions . . . "[103]

The most revealing proof of the quasi-religious function of technology is the fact, he observed, that even those who are ruined or deprived of their livelihoods by it, even those who are its critics and attackers, "have the bad [i.e., guilty] conscience of all iconoclasts."[104] We can only add that for those who belong to the ranks of technology worshippers today, questioning or criticizing technology is the most unforgivable form of heresy.

TECHNOLOGICAL CONTROL OF THOUGHT AND BEHAVIOR TO ENSURE CONFORMITY AND EFFICIENCY

The importance of the so-called "human techniques." Modern technological societies not only exist in a philosophical vacuum, derive acceptance from a quasi-religious worship of their technology, and legitimize themselves by means of their technological accomplishments, but they also engineer conformity and maximize their efficiency by using technological mechanisms of social control. Indeed, for the first time in history, the recourse to purely technical means of manipulating human beings is as important for the successful functioning and preservation of the existing social order as the technological control of nature itself.

This development is in no way surprising, since the vocation of all technology is to make everything happen according to predetermined objectives, rather than letting anything happen on its own. As this technological way of being in the world became the central ethos of modern society, it was only to be expected that human beings, the practitioners of technology, would increasingly be subjected to its control as well.

For his part, Ellul realized the importance of human-directed technologies in the new, technological society. He referred, instructively, to the rapid development and refinement in the twentieth century of what he called "the human techniques."[105] Marcuse, too, was sensitive to the importance of the technological control of humans in an advanced industrial order. Regarding work, he remarked, "domination is transfigured into [technical] administration."[106]

According to Marcuse, even leisure, the quintessential sector of individual autonomy in the past, is subject in advanced industrial society to external administration or management. He wrote of the "promotion of mindless leisure activities;"[107] and Ellul, for his part, spoke of the recourse to mass entertainment techniques to compensate the worker in a technological society for the "fatigue of technical labor."[108] To this end, he specifically mentioned the importance of the mass communications media, as well as the existence of family vacation camps where activities were organized so that "no one is ever left to himself, even for a moment."[109] We can only comment with regard to this remark that in today's postmodern society the ubiquitous presence of television in the home; radios, sound systems, and cellular phones in automobiles; and the cell phones, Walkmans and iPods in the possession of people on the street serve the same objective of co-opting time when the individual might otherwise be alone and able to reflect on his or her situation.

Of the so-called "human techniques," Ellul singled out propaganda, the software technical means for organizing the thinking, emotions, and behavior of

entire populations along predetermined lines, for a central role in the technological society. Not only did he argue that modern democracies are actually
technocracies where the real decisions about war, peace, finance, public health,
economics, work, energy, and transportation are made by non-elected expert
advisors and ministers at the highest levels of government; he also maintained
that propaganda is necessary to preserve an illusion of democracy by
manipulating the people into supporting, and even advocating, policy decisions
that they have had no meaningful role in adopting.[110] He maintained that the
only way to maximize the efficiency of each individual whenever he or she is
called upon to think, act, or believe is to use propaganda to provide myths and
promises that eliminate any questioning, hesitation, or doubt.[111] And he recognized the importance of television as a means for delivering propaganda, saying
a half-century ago that it was truly revolutionary because it reached individuals
in the privacy of their homes and influenced, without the slightest effort on their
part, their thinking and behavior.[112]

Skinner's paradigmatic behavior control technology utopia. Consistent
with the emphasis in a technological society on engineering maximal conformity
and efficient control, a new academic discipline, behavioral psychology, associated with the control of humans themselves and inspired by the ideas of J. B.
Watson, was included as a discipline in a number of American universities early
in the twentieth century. In 1948, B. F. Skinner, a leading behavioral
psychologist at Harvard, communicated some of its most ambitious claims to a
broader public in a novel, *Walden Two*, which sold over a million copies. In
1971, he published *Beyond Freedom and Dignity*, a theoretical exposition and
refinement of the ideas he had broached in *Walden Two* and a blueprint, in
effect, for an entire society whose members would be programmed by means of
behavior control technology to be happy and good.

No longer would citizens be left free to choose their own actions, Skinner
said, nor would their dignity be respected by giving them credit for what they
did, because in his utopia everyone would be programmed in advance to do the
right thing. A technique of administering rewards at strategic points to lock in
patterns of good behavior, developed by Skinner in experiments with animals
like rats and pigeons and then applied to humans, was the basis for his plan. The
ultimate goal, or good, he said, would be to engineer behavior that would promote the survival of the culture.

In support of his utopian scheme, Skinner argued that there was no innate
self, only a repertory of responses that had been conditioned by external stimuli.
Accordingly, he reasoned that human techniques designed to condition a person's behavior for the better were consistent with human nature and therefore
legitimate. He also maintained that his method was more effective and more
pleasant than traditional forms of behavior control like threats, persuasion, and

punishment, each of which had the additional disadvantage of leaving individuals free to make their own (often wrong) choices.

Revealing some similarity to Christian theologians who believed technological progress was to be part of the process of human recovery of qualities lost with the Fall, Skinner argued that we should use "physical, biological, and behavioral technologies" to "save ourselves" from our current problems and survive.[113] For him, these problems included pollution; resource depletion; rejection of school, work, and military service in Vietnam by the young; and the threat of nuclear war.[114] In linking his behavior control technology with a kind of salvation, he showed his fidelity to the tendency of humans, particularly in the West, to invest technology with their most extravagant expectations and expect that it will bring a magical transformation that will solve their most urgent problems.

Skinner's ambitious utopian proposal implicitly rejected the free choice of democracy, even as a convenient illusion, and explicitly abandoned the Enlightenment ideal of the autonomous individual. In effect, Skinner wanted to rely on a purely technological solution to the age-old political problem of regulating behavior and containing human conflict in a society. In making his choice for efficient order, he rejected the competing interest of personal freedom that political philosophers like Locke and the American founding fathers had factored into the political equation. Despite a professed self-identification with "humanism," Skinner's utopian proposal actually amounted to a betrayal of humanistic values and demonstrated the potential for social regimentation and dehumanization when humans are the objects of efficient technological control.

Marcuse's thesis of latent aggressivity in a social context of pervasive behavior control. Sharing with Ellul a distaste for the values implicit in a Skinnerian-type blueprint for total social control, Marcuse referred critically to the "administered living" that prevails in advanced industrial society. He derisively wrote of a "society of total mobilization," where the "established technical apparatus engulfs the public and private existence in all spheres of society."[115] In these societies, technical means of behavior control are already used to ensure that individuals satisfy the norms of thought and conduct. Little, therefore, is left to chance or to the autonomy of the individual. In addition to the likelihood of political oppression and workplace exploitation at the hands of those possessing the technical means to engineer conformity and productivity in such societies, Marcuse warned that there was an instinctual price to pay.

Applying the Freudian tools of psychological assessment included in critical theory, Marcuse focused on the dynamics of the life instincts, eros and aggressiveness, in advanced industrial society. Indeed, if this instinctual factor has the importance Marcuse ascribed to it, the technological successes of advanced industrial, or even postindustrial, society may not be enough to confirm its validity nor prevent it from being recognized as a form of social pathology.

In *Eros and Civilization* (1966), Marcuse argued that the erotic component of instinctual energy—the source of love, friendship, and social harmony—is actually drained and weakened by alienated labor performed in industrial production. At the same time, he warned, there is a build-up of hostility because of the anger and frustration caused by the total administration of people's lives, largely achieved by behavioral techniques. That is, individuals feel hostile toward a depersonalized source of authority that, in place of the traditional role exercised by the parent in the framework of the family, operates through the schools, the media, and the workplace to control behavior. There is, he said, a "whole system of extra-familial agents and agencies" involved in this ongoing process of manipulation and control.[116]

Since individuals in a society where sophisticated techniques of behavior control are used cannot readily identify a specific authority figure like a parent responsible for their manipulation and control, Marcuse argued that "aggressive impulses empty into a void."[117] The result is not really an "emptying," but, as Marcuse explained, a buildup of aggressive tension incapable of identifying its real target and turning inward, therefore, toward the ego itself. Guilt and neurotic tendencies are thus aggravated, increasing the likelihood of external outbreaks of persecution and violence. Furthermore, the aggression problem is compounded because some of the instinctual energy of eros that normally counterbalances it is absorbed in alienated labor. When a society reaches such a point, Marcuse warned, the expanded reservoir of aggressive energy can be tapped into by leaders or demagogues and directed at scapegoats—that is, toward "the omnipresent but hidden foe," he said, "whose presence" requires "total mobilization."[118]

When Marcuse wrote these words, the hidden enemy within was often characterized as a "communist" or "fellow traveler" (i.e., a communist sympathizer and collaborator during the Cold War). A contemporary equivalent to the communist foe who required total mobilization then is today's "sleeper," the immigrant who ostensibly leads a normal life in America for an extended period but should warrant our mistrust and hostility in President George W. Bush's "war on terror" because he may eventually commit a terrorist act. Thus, as Marcuse explained, in any culture where there is a pervasive apparatus and sophisticated array of techniques of social control, the problem of the accumulation of aggressive energy that threatens to burst through the surface of the social order is indeed a reality.

Postmodern Technoculture: Existence without Philosophical Direction and Technology without Limits

THE GENESIS OF DERRIDA'S DECONSTRUCTIONIST PASSPORT TO THE POSTMODERN VOID

The proximate philosophical developments that provided the broader ideological context for the emergence of a new, postmodern brand of thought took place in post–World War II France, where Jean-Paul Sartre initially played a leading role. Although he had been influenced by Nietzsche's rejection of metaphysics and conventional ethics, Sartre advocated an active human stance in the world that had nothing in common with the analytical philosophers' trivialization of philosophy as linguistic analysis or with Heidegger's emphasis on a passive human "shepherding" of Being. In a sense, we could say, Sartre fought a rear guard action against the cultural anomie and philosophical chaos that postmodernism was ultimately to bring. His thought, despite its existentialist affinities with Heidegger's philosophy, retained elements of the Enlightenment grand narrative of progress and brought a left-leaning political and humanistic focus to French philosophy in the early postwar period.

In his famous essay, *Existentialism is a Humanism* (1945), Sartre emphasized the idea of human freedom in a world where God, predetermined essences, and transcendent truths conveying higher meaning were no longer viable conceptions. Accordingly, he assigned to men and women the task of actively shaping their own being as humans and determining the significance of the world around them. In contrast to Nietzsche's call for an elite of overmen who would practice an "authentic will to power," however, Sartre emphasized commitment to progressive social change and humanistic responsibility to others. It was precisely what Heidegger termed Sartre's "anthropologizing of Being," however, that aroused strong objections on Heidegger's part.[119]

When he read Sartre's essay, which he interpreted as a travesty of his own existential philosophical position, Heidegger responded with his polemical *Letter on Humanism* (1946). There, in the aftermath of the German defeat, he reiterated his view that all the problems of modern culture could be traced back to Plato's metaphysics, which had placed humans at the center of creation and led to a continual, misguided "anthropologizing of Being." The only way to rectify this fatal error, Heidegger argued, was to dismantle the metaphysical tradition (i.e., *Destruktion*).[120] This declaration amounted to a significant radicalization of his earlier position, a radicalization that Jacques Derrida and a new generation of postmodern thinkers in France eventually sought to carry even further.

In contrast to Sartre, a number of important French intellectuals in the early postwar period did break with the humanist tradition and did reject the philosophical narrative of progress. One of the most influential of these was Claude Lévi-Strauss, an anthropologist who wrote in *Tristes Tropiques* (1955), his famous account of field studies among tribal peoples of the Amazon in the 1930s, "The world began without the human race, and it will end without it."[121] Originator of the new structuralist methodology, by which he applied the theoretical concepts of linguistic structure to a study of society as a whole, he argued that societies were structures with relatively stable relations among elements that followed no rational pattern of historical development.[122]

His structuralist theory also amounted to a rejection of the Eurocentric view of humanity, according to which Enlightenment thinkers and their successors had enunciated universal truths about the human condition. Lévi-Strauss, in contrast, recognized each culture as autonomous and worthy of respect in its difference or otherness.[123] Thus, in his investigations, he found neither a universal human nor a universal history. One of his most prestigious structuralist successors, Michel Foucault, even considered the human to be a mere "site" where cultural, social, economic, and psychological forces fortuitously interacted. In 1966, Foucault declared that "man" was a recent invention that would soon disappear, like a face drawn in the sand.[124]

In 1968, as a student-led rebellion against the regime of General de Gaulle was going on in the streets of Paris, Derrida thrust his way to the forefront of the French philosophical scene when he delivered his lecture, "The Ends of Man," at a colloquium in New York. He shared with Lévi-Strauss and Foucault their rupture with humanism and the ideology of progress, but he was the philosopher who most pointedly identified himself in France with a post-Heideggerian project to dismantle the metaphysical tradition.

Despite his Jewish roots, Derrida took Heidegger, the German philosopher who exhorted the German people in 1933 to be open to the "Promethean theory" of National Socialism, as the starting point for his new philosophy. Heidegger had called in 1946 for the destruction of the metaphysical tradition as a response to the "anthropologizing of Being," but Derrida decided to adopt an even more radical project. He accused Heidegger of inconsistently returning to humanism by making man the "shepherd of Being." And, in what could perhaps be characterized as a nihilistic undertaking, he called for the deconstruction of the language of philosophy itself.[125]

For Derrida, the philosopher's task would be to make a deconstructive analysis of the logos of any given text, to show its paradoxes and inconsistencies, revealing the lack of connection between its words and intentionality. Since any text contains ambiguities and can be read in different ways, Derrida argued, no definitive interpretation was possible. No longer would the author be placed above the reader, nor would the meaning (i.e., the signified) be given greater importance than the written text itself (i.e., the signifier).[126]

This objective was in contrast to the analytical philosophers, who believed the proper task of philosophy was to clarify or describe the use of language. Derrida's project went beyond the loss of credibility of the traditional philosophical narratives, rather, in an effort to show there was an absence of rationality at the core of the texts themselves. It contributed to the establishment of the new, amorphous, and anarchical postmodern cultural framework in which Jean-François Lyotard's *Postmodern Condition: A Report on Knowledge* assigned a crucial role to information technology.

It is ironic, perhaps, that Derrida, derided in France by some of his colleagues as representative of an "antihumanist" current of the 1970s and never honored with a prestigious post in a French university, produced works that, despite their esoteric and technical nature, were translated into over fifty languages. And it is significant, perhaps, that he enjoyed his greatest success and notoriety in the United States, the Western nation most advanced in technology and among the least inclined to philosophy. There, according to a critical assessment by a journalist in a leading French weekly at the time of his death in 2004, he had been received in academic circles as a Parisian "cultural hero" and treated as if he were "the greatest philosopher in the world."[127]

LYOTARD'S FORMULATION OF A LEGITIMIZING PRINCIPLE FOR POSTMODERN SCIENCE AND CULTURE

Wittgenstein's post-Nietzschean focus on "language games" as his starting point. According to Lyotard's analysis in *The Postmodern Condition: A Report on Knowledge* (1979), there were two philosophical "grand narratives" that had provided the basis of legitimization for scientific knowledge before the collapse of metaphysics and higher values that Nietzsche announced at the end of the nineteenth century. These were the Enlightenment theory of progress, according to which scientific knowledge would serve the pacification of nature and provide a basis for human autonomy and freedom;[128] and German idealism of the nineteenth century, which encouraged a "speculative spirit" and justified the acquisition of scientific knowledge as an end in itself.[129]

With the loss of these philosophical narratives, Lyotard said, there was a "fall into philosophical pragmatism" (i.e., positivism), which brought what he called the "performativity standard" to the forefront as the only remaining value that could serve a legitimizing role.[130] That is, the pragmatic fact that scientific knowledge provided operational power when technologically applied, rather than the existence of a credible philosophical narrative guaranteeing that it was a way to the truth, was accepted as a basis for legitimacy.[131] This is essentially what Marcuse said, as we have seen, about the legitimizing role of technology in advanced industrial society.

By the 1960s, however, the transition to postindustrial society and post-modern culture was, in Lyotard's eyes, under way; and the reliance on the pos-itivistic principle of performativity for legitimization turned out to be what he qualified as only an "episode."[132] To explain the most important philosophical developments underlying this new transition, Lyotard concentrated on the post-Nietzschean focus on language. Singling out Wittgenstein for special attention, Lyotard stated that unlike the other philosophers associated with the Vienna Circle of logical positivists, Wittgenstein eventually had gone beyond positive-ism. In effect, he had shifted his emphasis from the idea that only the methods of natural science could provide knowledge of reality to the idea that the proper role of philosophy was to analyze language and clear up the ambiguities and misunderstandings it had created. Or, as Lyotard chose to put it, Wittgenstein had found a new way to legitimize knowledge with a focus on its rules of language discourse (i.e., syntactic and semantic properties) after the philosoph-ical narratives were lost.[133]

At one point, Wittgenstein declared that philosophy was to be "a battle against the bewitchment of our intelligence by means of language," and he adopted a project "to bring words back from their metaphysical to their everyday use."[134] He soon realized, however, as Lyotard pointed out, that there was no one, universally valid metalanguage or set of rules for language. This was in contrast to Bertrand Russell, the most famous of those analytical philosophers who had searched for a perfect, or ideal, metalanguage that would fit all occa-sions.[135] Rather, explained Lyotard, Wittgenstein understood that there was an indeterminate number of language "games," each with its own rules of discourse corresponding to its particular area of knowledge.[136] To describe this "splin-tering" in language games, which he believed also applied to postmodern culture as a whole, Lyotard employed Wittgenstein's metaphor of "an ancient city: a maze of little streets and squares, of old and new houses, and of houses with additions from various periods, and this surrounded by a multitude of new boroughs . . ."[137]

The fact that this splintering had left the discourse of science with its own epistemological language had already started to become clear, said Lyotard, when Kant separated theoretical reason (*Vernuft*) and practical reason (*Verstand*).[138] Furthermore, Lyotard pointed out that there was even a "plurality of languages" within the separate boundaries of the various areas of scientific knowledge itself.[139]

Recognition of paralogy, or subversion of existing norms in search of the new, as the defining language game of postmodern science. Keeping the foregoing considerations in mind, Lyotard turned his attention to the rules of the language games associated with the pragmatics of science. Before postmodern-ism, he stated, the emphasis was on the "denotive" games that governed scien-tific discussions of knowledge and indicated what was true or false with respect to particular referents. The postmodern development, however, brought to the forefront examination of the "prescriptive," or action, language games that determine the rules (i.e., "metaprescriptives") by which scientific knowledge was deemed to be valid in the first place.[140]

Citing in 1979 recent developments in quantum mechanics and microphysics and the influence of mathematicians like René Thom, Lyotard asserted that the pragmatics of science had taken a new, postmodern direction. Thom's catastrophe theory had shown mathematically, he said, that under certain local conditions, determined phenomenon are subject to discontinuity and disruption. Such a demonstration meant, in effect, that the assumption of stability in natural systems, upon which the principle of determinism and the standard of performativity were predicated, was untenable.[141] Postmodern science, Lyotard therefore concluded, had made a shift and was "theorizing its own evolution as discontinuous, catastrophic, nonrectifiable, and paradoxical."[142] Scientists, he said, no longer seek agreement within the existing logical framework of science but try, rather, to undermine it from within.[143]

This paralogical project in science has an obvious affinity with Derrida's project to deconstruct philosophical texts. His aim in "neutralizing communication," Derrida said, was "to provoke . . . a new tremor or . . . shock . . . that opens a new space of experience."[144] In the case of postmodern science, Lyotard explained that the goal of paralogy was to produce the unknown, not the known. The emphasis was no longer on performativity but on making a new "move," to open the way toward new epistemological rules circumscribing a new field of research for the language of science.[145] Thus, at this point Lyotard joined Derrida in what essentially was a common purpose—the undermining of existing truth in order to move on to the new, presented as an end in itself. It is a process, in effect, whose real goal is only to continue.

Designation of paralogy as the legitimizing paradigm for a postmodern culture of fragmented and temporary knowledge. In Lyotard's view, although the "grand [philosophical] narratives" are absent from postmodern culture, "little narratives" remain the quintessential form of imaginative invention, particularly in science.[146] He pointed out that there is a multitude of these "little narratives," including a multiplicity of arguments concerning the rules (i.e., "metaprescriptives") for the validity of such knowledge. "Limited in time and space," he explained, their characteristics of locality and temporality correspond to the current "evolution of social interaction . . . where the temporary contract is in practice supplementing permanent institutions."[147] Furthermore, he found the same splintering of language rules in the pragmatics of social discourse as in the discussion of scientific knowledge. It was impossible, he said, to find a general metalanguage to cover "the totality of statements circulating in the social collectivity," just as there was no scientific metalanguage in which all other languages could be transcribed or evaluated.[148]

Despite, or perhaps because of, these difficulties, Lyotard proposed using the paralogical paradigm for the pragmatics of science as the legitimizing basis for knowledge in postmodern society as a whole. In doing so, he emphasized that he was proposing an "open model" for society. That is, a postmodern cultural establishment should not attempt to impose an orthodoxy; and a linguistic statement in any given area of knowledge should be considered valid if it generated ideas. The traditional ideal of a scientific, and social, consensus, therefore, would be recognized as an outmoded goal. That is, any consensus achieved would be recognized as only local and limited in time and space in the sense that it would be accepted by the players presently involved but acknowledged as subject to future termination.[149] In other words, the cultural verities of postmodern society, essentially the product of "little narratives," would be only those of the transient here and now.

Fredric Jameson, the Marxist scholar who wrote a critical introduction to the English translation of Lyotard's essay on postmodern knowledge, described this application to society of the principle of paralogy, with its elements of "innovation, change, break, renewal," as an attempt to lend to an otherwise repressive social system a rejection of conformity and to kindle the "disalienating excitement of the new and the unknown." Jameson found this effort unoriginal, however, nothing more than a transfer of the ethos of innovation of modern art to science and research. Lyotard's postmodern emphasis on a "dynamic of perpetual change," he added, boiled down to promotion of what is essentially a basic feature of capitalist society, as Marx had critically pointed out in his *Manifesto* one hundred and thirty years earlier.[150]

While in no way taking issue with Jameson's conclusions, we would like to add that Lyotard's essay is also evidence of the philosophical void that lies at the core of contemporary postmodern culture. What Lyotard identified as its defining ethos is the same restless need of humans to be continually acting and innovating that Pascal and Schopenhauer cited in their philosophies. The difference is that this tendency, attributed by Pascal to human nature and Schopenhauer to will, has been elevated by Lyotard to the status of a guiding cultural principle without any regard for its nihilistic implications.

THE CENTRAL ROLE OF INFORMATION TECHNOLOGY IN POSTMODERN CULTURE

In a postmodern context in which fragmented, transitory truths are recognized as the only possibility, Lyotard argued, information technology has a practical role of primary importance. Computers, he stated, supply information that would otherwise be lacking to the various groups discussing meta-prescriptives (i.e., language rules) for their branches of knowledge. Such information, in other words, continually feeds the never-ending paralogical process of postmodern science, which, absent the grand philosophical narratives, is the source of its legitimization.

The public, therefore, was to be given free access to memory and data banks in Lyotard's idealized version of postmodern society; and discussion would then be based on what he called "perfect information." There would be no permanent winners or losers, he added, because an inexhaustible reserve of knowledge (i.e., language's endless stock of possible utterances) would ensure that discussion would never reach a terminal point. The resulting politics, he concluded, would therefore serve both the need for justice and the pursuit of the unknown.[151]

In *The Coming of Post-Industrial Society* (1973), Daniel Bell had emphasized the importance of computers as a tool for the "intellectual technologies" (i.e., managerial techniques) upon which postindustrial (i.e., technological) society depended. These intellectual technologies, of which systems analysis was "the most ambitious version," according to Bell, were capable of taking into account numerous variables, assessing risks and probabilities, and making choices on the basis of algorithms rather than intuition. With computers, Bell said, it was possible to deal with multiple variables at the same time, to link economic theory with empirical data, and to develop econometric models for forecasting.[152] In other words, computers were a necessary tool for decision making and for engineering results in the complex governmental and corporate organizations that characterized postindustrial societies.

Lyotard's essay on postmodern knowledge appeared just six years after Bell's *Post-Industrial Society*; but it is important to note that in the context of emerging postmodern culture, he assigned to computers an even more fundamental role: the acquisition of knowledge itself. And because they transcend the capacity of their users, he added, computerized data banks are the "*Encyclopedia* of tomorrow."[153] This was a direct reference to the important role Diderot's *Encyclopedia* played in eighteenth-century France for disseminating the new scientific, technological, and philosophical knowledge of the time essential for progress. Data banks are, in Lyotard's words, "nature for postmodern man,"[154] meaning they are a vast, rich resource for humanity because they provide information.

Andrew Feenberg has appropriately remarked that at the time Lyotard published his essay, his analysis of the computer was "speculative" and his understanding of computers was "rather limited."[155] Furthermore, Lyotard was perhaps somewhat careless in equating the contribution of the *information* available in computers with that of the *knowledge* contained in the Enlightenment *Encyclopedia*. In making his ambitious claim for data banks, in any case, Lyotard demonstrated a fidelity to the Western ideological tradition, born in the Middle Ages, of linking technological advances to the realization of the most profound human purposes.

Another virtue of computers for Lyotard was that in mechanizing and depersonalizing knowing into a kind of social function, they undermine the humanistic ideal of knowledge as the "self-construction" of the individual subject.[156] In accepting their exteriorization and separation of knowledge from the individual knower, he showed his affinity with other antihumanist, post-Nietzschean philosophers in France like Lévi-Strauss, Foucault, and Derrida. They, too, rejected the idea of a human subject whose reason was capable of dictating universal truths, an exercise which Heidegger had characterized as the unfortunate "anthropologizing of Being."

Consistent with Lyotard's buoyant rhetoric, some technologists, politicians, and corporate entrepreneurs and publicists enthusiastically spoke in the last decades of the twentieth century of the new Information Age and promoted the computer as the practical basis for a new, more advanced form of society and culture. In addition, they treated the acquisition and processing of information, as Lyotard suggested in his description of computers as supplying an "inexhaustible" reserve of information that would facilitate a never-ending search for the "unknown,"[157] as a new value or end in itself.

POSTMAN'S THESIS THAT INFORMATION CHAOS ON THE WEB PRODUCES A CULTURE WITHOUT A VIABLE MORAL COMPASS

Lyotard's articulation of his vision of a computerized postmodern society coincided with the groundbreaking inaugurations of the Prestel videotex system in England in the late seventies and the Teltel information net in France in 1981.[158] These experiments in computer-mediated communication (CMC) networks with centralized data banks were followed by an exponential broadening of the base of personal computer ownership in a number of affluent nations and the opening by the Clinton Administration in the mid-nineties of Internet access to the general public. (Access had previously been restricted to specific academic, corporate, and governmental users.) At this point, the high expectations of Lyotard and others about the benefits of information technology were energized anew.

The Internet was, indeed, a quantum leap in information technology. Not only do public and private data banks exist today, but, thanks to the Web, there is an extensive global network of computerized interpersonal communication. Such changes have made it possible for any individual logged onto the Web to communicate with anyone else who is logged on worldwide. Despite the fact that the overwhelming majority of exchanges between users are trivial in terms of expanding knowledge, there are undoubtedly some that contribute to the "little narratives" about knowledge that, according to Lyotard, characterize postmodern culture.

Because of the nature of the Internet, however, a kind of libertarian anarchism of narrow, often mutually exclusive, interests of a vast multiplicity of individuals and groups, characterized by its promoters as an ideal, essentially reigns. Although the Marxist critic Terry Eagleton qualified the cultural perspective of postmodern deconstructionism as "libertarian pessimism,"[159] it would be more accurate, perhaps, to describe it as libertarian nihilism. This terminology not only expresses the emptiness and confusion of postmodern technoculture, but it characterizes as well the philosophical void which underlies the fragmented and chaotic information universe of the Internet. In such a cultural context, there are no accepted higher values upon which one could ground a viable philosophical challenge to the promoters and practitioners of technology without limits.

Postman, who brought a necessary critical perspective to the cultural impact of information technology, called the present situation Technopoly, where traditional philosophical narratives have disappeared from the horizon, technical values reign supreme, and information technology is the decisive element.[160] When he coined this term in 1993, however, he was quick to add that the United States was the only country where this cultural transformation had been completed.[161] It is open to question whether it can now be said, a decade later, that it has been completed in other societies where traditions go back many centuries more, even millennia more, and where personal access to computers and the Internet is significantly less generalized.

Like Lyotard, Postman recognized that the computer has acquired a crucial role in this new social and cultural system, which is postmodern and technopolistic. A computer is, he said, the "quintessential ... machine for Technopoly," providing the dominant metaphors for society. That is, such thinking redefines humans as "information processors" and nature as "information to be processed."[162]

Furthermore, he noted that the computer itself "is almost all process."[163] Indeed, if we turn our attention to a computer's content, much of it is of a very superficial and time-specific nature, intended for immediate use. It is not at all surprising, therefore, that the computer has become a particularly valuable instrument for such transitory activities as military offensives, speculation in financial markets, and making travel and hotel reservations for business trips or vacations.

One dictionary definition of the word "information" is "unorganized or unrelated facts or data." The computer serves to store and manipulate, or process, this information, which has usually been extracted from its original context and broken into parts—essentially, commodified—for purposes of profit or control. This truncated data, which in many cases no one is even interested in assembling to provide the basis for real knowledge, is the computer's real "content."

As Postman accurately pointed out, although the computer is an intelligent machine, it is incapable of telling us what statements really *mean*. That is, a computer can correctly interpret the logic of statements; but lacking any feelings or real-life experience, it cannot understand what their deeper meaning, if any, is.[164] Thus it is not surprising that in the Information Age of computer dependency, so many humans appear to know so little.

The computer revolution, according to Postman, brings us new forms of information in unprecedented amounts and at increasing speeds of delivery.[165] This has resulted, he said, in a new and serious problem, the creation of a glut of information. A climate of information chaos now prevails, in his view, where the link between valid human purposes and information has been severed.[166] Technopoly therefore suffers from a kind of cultural AIDS, Postman asserted, because its "information immune system" has been overwhelmed by a plethora of information.[167] Normally, social institutions like the family, political parties, universities, religion, and the state served to filter and control the spread of information; but in the information chaos of Technopoly, he said, they are unable to retain their effectiveness.[168] Information, therefore, has become a form of garbage, incapable on its own of providing any positive response to the most fundamental human questions.[169] The overall effect is a loss of meaning, a general breakdown in psychic tranquility and social purpose.[170]

It is precisely this uncontrolled inundation of information that has made information technology, along with television, an instrument of destruction of culture and traditional values. For this reason, it is not strictly accurate today to refer to "postmodern culture" or a "technoculture." In effect, not only have humans and nature been redefined in the Information Age; but culture itself has been fatally undermined. True culture serves the purpose of giving a people, both collectively and individually, a meaningful place in the world, an historical continuity in their experience, and an enduring positive direction, beyond the transitory rewards of conquest and material riches. It creates values and generates ideals, drawing its sustenance not only from factual knowledge, but also from art and myth, which are products of imagination, not information. This is why the danger of what Lyotard called a "reduction to barbarity" with the loss of the "grand narratives"[171] has not really been overcome. And this is why, in the present postmodern ideological context, nihilism threatens and technology without limits thrives.

The consequences of this situation are magnified by the fact that information technology, along with other electronic media like television, produce a loss of physical reality as individual consciousness is absorbed into the virtual reality of cyberspace. As humans who have lost their cultural points of reference experience increasing alienation of consciousness from external reality, the likelihood of a catastrophe without precedent increases as the power and destructive capability of technology grows. Needless to say, this alchemy of magnified technological means, cultural anomie, and alienation from the real world in which we live produces a mix of the most unstable and dangerous kind.

The dead bodies in American schoolyards of students shot in the late 1990s by fellow classmates armed with grenades and semiautomatic weapons and primed for violence by violent fantasy material they watched on information media provided a disturbing example of the impotence of socializing institutions like school and the family in the face of uncontrollable information and the alienation from real life of those who become habituated to the alluring virtual pastures of videogames and cyberspace. Harbinger in a strange way of the 2001 terrorist attacks on New York City and Washington, D.C., the 1999 Columbine, Colorado, high school massacre was also carried out by suicide attackers. Whereas the 9/11 jetliner hijackings were largely motivated by opposition to U.S. foreign policy in the Middle East, however, the Colorado massacre appears to have been an expression of pure nihilism, a tragic product of the adolescent American experience of growing up in cyberspace and in a postmodern culture empty of any valid higher meaning.

BIOTECHNOLOGY AS THE VERY ESSENCE OF POSTMODERN TECHNOSCIENCE WITHOUT LIMITS

The crucial practical contribution of information technology to biotechnology. The 1973 method for recombining DNA made it possible for scientists to add genetic characteristics to living organisms and thus bioengineer, in the laboratory, modified forms of life. This technology, indeed, has the most profound implications for humanity. It has the potential of leading, as Jeremy Rifkin put it, to a "second Genesis," by which humans would deliberately combine traits of different species and thus create entirely new forms of life,[172] presumably to satisfy their own needs and objectives. There is a danger, however, that the broader and unintended consequences of such endeavors, in a human as well as an ecological sense, may be both harmful and irreversible.

In the early 1950s, two important events laid the groundwork for this biotechnological revolution. First, Francis Crick and James Watson discovered that the genes that are the basis for specific traits and for the development of living organisms are composed of double-helix DNA molecules. Second, the U.S. Census Bureau installed the first working electronic computer in Blue Bell, Pennsylvania. According to Rifkin, it was the subsequent fusion of the computer and biology that signaled the beginning of a new age of biotechnology.[173]

Information technology, indeed, had a decisive impact on the development of biotechnology on a practical level. Computers made it possible to store, analyze, and classify the vast amounts of data necessary for the identification, mapping, and sequencing of the genes of living creatures. How else could scientists compile a data base of three billion entries for the approximately 20,000 to 30,000 genes[174] that comprise the human genome and determine the variable sequence of the thousands of DNA nucleotide components that make up each gene and are necessary to identify it?[175]

Subversion of limits in biotechnology by new, theoretical paradigms from cybernetics. Equally important for biotechnology was the influence of the principles of computer technology on biological theory. History, as we have seen in the case of Newton's clockwork model of the universe, is not devoid of examples applying technological models to knowledge of a higher order. By applying the cybernetic principles of computers to organic life, biologists adopted a new paradigm for the functioning of living systems. Engineering theory, in effect, became the new basis for explaining nature.

Cybernetics is a theory of external information absorption, storage, and processing, including self-regulating feedback capability, all subject, according to biologists and geneticists, to internal, DNA-coded instructions in the genes. They thus understand the functioning of all living organisms as self-programmed activity, including, under extreme conditions of energy fluctuation within them, an evolutionary possibility of reorganizing their structures at a higher level of complexity.[176] This functional, information-processing paradigm not only corresponds to the operation of human-made mechanisms (computers); it also has the advantage, in an age of technology without limits, of inviting human manipulation of natural organisms.

The cybernetic model of living organisms also permits scientists and technologists seeking their modification to replace the idea that they were discrete structures with a permanent set of attributes by the theory that they are merely "bundles of information." This means that these biotechnicians and genetic engineers no longer recognize the idea of fixed biological borders as a limit. Matter, or structure, is redefined as energy—and energy as information, subject to constant modification.[177]

This transformation in biological thinking amounts, in Rifkin's view, to a new and more extreme desacralization of nature. It means, in effect, that existing ethical limits on human tampering with the process of life have been discarded and humans can now modify the genetic raw material of life without due moral concern. Biotechnicians have already created transgenic animals by transferring, without respect for existing species boundaries, the genetic characteristics of one plant or animal species to another, animal traits to plants and vice versa, and human traits to animals.

Furthermore, in an assessment that has its similarities to Lyotard's emphasis on the process of information acquisition in postmodern culture and its relation to the production of knowledge, biological scientists consider evolution itself as a process involving a series of creative advances by living organisms that bring them improvement in this accumulation and processing of information.[178] In this sense, they claim that each new species is better informed than its predecessor;[179] and see no end to this process of improvement because, as conditions continually change over time, "information gathering" continues.[180] Since, in the workings of nature, the "new" in an organism comes as a response to information from the outside, these scientists argue that the bioengineering of living organisms by splicing into them DNA coded with new information is in conformity with the workings of nature.[181]

Thus, in a postmodern technoculture where metaphysics and ethics texts have been deconstructed, where scientists are encouraged to engage in a paralogical process of destruction of existing rules for determining truth, and where acceptable knowledge is fragmented and specific only to time and place, organic life and the evolutionary process itself have been redefined to legitimize an open-ended project to transform living species, ultimately subject to the vicissitudes of human will. It is a culture, in effect, where whatever is technologically possible is permissible.

The commercial and scientific contradictions of biotechnology. In the present context of capitalist economics, many biotechnologists will do whatever appears in the eyes of corporate entrepreneurs to be commercially feasible. In 1980, the U.S. Supreme Court ruled that bioengineered organisms could be patented and sold as commercial products. At that fatal point, the Faustian power of biotechnology, armed with the new scientific theory borrowed from information technology, was allied with the dynamism of the profit calculus, which subordinates everything to a never-ending process of capital accumulation.

Rifkin theorized in 1983, somewhat charitably perhaps, that a broad biotechnological project to manipulate the genes of nonhuman forms of life would be based on a desire to enhance the "survival prospects" of the human species.[182] Presumably, such applications would include modifications of food crops, innovations in animal husbandry, enhanced production of organic resources like trees and fibrous plants, and environmental cleanup. In reality, however, it is clear that the overriding motives for biotechnological manipulations such as these are—and undoubtedly will be—monetary profit and short-term convenience. For example, biotechnicians working on agricultural applications have transferred flounder genes to fruit and vegetables, moth genes to tomatoes, mouse genes to tobacco, and human genes to pigs.[183] In each instance, the modification was performed for the sole purpose of giving the host organism a commercially advantageous characteristic, such as increased shelf life, prolonged resistance to spoilage when refrigerated, or accelerated and enhanced growth.

There is, indeed, good reason to be skeptical, on a purely scientific basis, about the long-term human survival benefits of relying on genetically-altered organisms. An organism that has been bioengineered by scientists to possess a marketable trait useful in an artificial and protected setting like a cattle feedlot or a cultivated farm plot, for example, may turn out to have a harmful impact when released in the greater complexity of a natural ecosystem. A case in point is an enzyme bioengineered for use by forest and energy industries to destroy lignin, the substance that makes wood rigid. If it migrated offsite, it could decimate millions of acres of forest.[184] Furthermore, artificially altering the DNA of existing organisms to add certain desirable characteristics usually results in the creation of undesirable side effects that render reliance on them risky. For example, "superpigs" that have been genetically altered to produce more meat by the addition of human growth hormones to pig fetuses are arthritic, cross-eyed, and afflicted by muscle degeneration.[185] Because of these weaknesses, their health and evolutionary capability are inferior to those of a genetically unaltered variety of the same pig. What these examples illustrate is that ignoring the question of limits in the case of biotechnology, regardless of the short-term profits and convenience, is to invite problems of a most serious nature for both humans and the artificial forms of life they have created.

Professor Sloterdijk's controversial proposal to bioengineer "improved" humans. Marx and Engels promised that when humans had conquered scarcity by means of advanced industrial technology and socialist distribution of goods, "truly human history" would begin and humans would pursue their "all-around development." It is also possible, however, that if humans considered the external environment to be "mastered," they would *not* enjoy enhanced autonomy and freedom. On the contrary, they, themselves, might be the object of a degree of technological manipulation and control surpassing all previous limits. This is the scenario, unfortunately, that our recent history and experience with technology suggest will be the more likely reality.

In the summer of 1999, a German professor of philosophy, Dr. Peter Sloterdijk, created a stir in German media and intellectual circles when he pointedly reaffirmed the Heideggerian thesis that "humanism was dead" and openly called for the convocation of a general assembly of behavioral scientists to draw up a "code" of guidelines for the future bioengineering of human beings—that is, for what he termed the "breeding" and "autodomestication" of humans. With this proposal, we encounter the recurrent belief that new technology will rescue humanity at a crucial moment and positively transform its destiny in a radical way. What was needed as a response to the failure of humanism, Sloterdijk argued, was a remaking of human characteristics guided by a predetermined plan. In the future, he said, civilization must consist of humans shaped in accordance with a new "anthropological technology."[186] His emphasis, however, was not on eliminating disorders or abnormalities, but on bioengineering humans for better quality.

It is significant that Sloterdijk, identified by the French daily, *Le Monde*, as a "postmodern philosopher," made his proposal in Bavaria at a colloquium on Heidegger. The title of his paper was "A Response to the [Heidegger's] *Letter on Humanism*." His postmodern response, however, as we have seen, was not really philosophical, but technological. One critic, writing in the German weekly *Die Zeit*, sarcastically qualified Sloterdijk's proposal as "The Zarathustra Project." The magazine *Der Spiegel*, which likewise alluded to the Nietzschean overtones of his project, accused him of advocating the creation of a new overman and labeled his ideas as "fascist rhetoric."[187]

If Sloterdijk's proposal were implemented by a nation-state in accordance with a eugenics policy, these criticisms would certainly be appropriate. Indeed, one may recall Hitler's *Lebensborn* program, based on recreation camps for young German soldiers, who were encouraged before they left for the World War II front to mate for racial quality with young women from Germany, Poland, and Norway, selected for their Aryan characteristics.

A less controversial application of biotechnology to humans would be essentially corrective: that is, the elimination of all genetically caused diseases, physical abnormalities, and antisocial tendencies.[188] Pursuit of this goal, however, could turn out to be counterproductive. That is, a project to "perfect" the human species by eliminating certain genetically borne defects would, if successful, reduce the diversity of the human gene pool in ways that might turn out to be detrimental in the face of future, altered environmental contingencies. For example, the sickle-cell anemia genetic trait is not only the cause of a disease but also a factor that increases the chances for survival of individuals stricken with malaria. As the world's climate becomes warmer and more humid in some areas, the presence of this trait in the human gene pool could prove, on balance, to be beneficial or even essential for survival.

Furthermore, the fact that the human genome was found in 2001 to include a total of 25,000–30,000 genes presents a serious obstacle to the success of Sloterdijk's "anthropological technology" proposal or to any attempt to bio-engineer human genes for whatever reason. Since the genes send messages to control the production of the 142,000 proteins present in the human body, the significantly lower total number means that each gene has multiple functions.[189] If biotechnicians do not discover a way to overcome this difficulty, tampering with a gene for one purpose will entail the risk of producing serious, unwanted, and unpredictable side effects.[190]

Taking account of the new genetic technologies, Paul Virilio, a contemporary French philosopher, has pointedly warned that humans may now be "created" rather than "procreated." In contrast to the "experiments on humans" performed by Nazi scientists like Dr. Mengele at Auschwitz during World War II, Virilio declared, there is a danger today that there will be "human experiments."[191] Indeed, the fact that some of the results of genetic manipulations performed on human embryos may be neither predictable nor desirable raises ethical questions of the most disturbing nature.

Undaunted, in any case, by the revelation in 2001 of the discrepancy between the number of genes and the greater quantity of proteins they produce to guide the development of a human organism, and despite the skepticism in the scientific community, some scientists and biotechnologists promptly announced their intention to push ahead by shifting their focus to what they call "proteomics." Accordingly, three technology companies, Oracle, Hitachi, and Myriad Genetics, revealed they were joining in a new, five-billion-dollar project to identify all the human proteins and their interactions within three years.[192] Thus, in one of the more egregious and recent examples, the bearers of the spirit of technology without limits reiterated their single-minded determination to press ahead.

VIRILIO'S RADICAL CRITIQUE OF POSTMODERN TECHNOCULTURE AND TECHNOLOGY WITHOUT LIMITS

The absence of ethical limits in postmodern technoscience. Writing at the end of the second Christian millennium and with the power and attendant risks of nuclear, genetic, and information technologies in mind, the French philosopher Paul Virilio moved the technology critique, in *The Information Bomb* (2001) and *Crepuscular Dawn* (2002), to a more radical level, punctuated with apocalyptic premonitions and ultimately presented as a warning of impending disaster on a grand scale—that is, what he called the "Total Accident." His escalation of the rhetoric, and substance, of the technology critique is not only a reflection of the quantum leap in the power and sophistication of recent innovations, but also a consequence of the exponential acceleration of the pace of contemporary technological development. Virilio himself expressed his agreement with Daniel Halevy's theory of the "acceleration of history," adding that with the emergence of cybernetics, all reality is accelerating, whether "things, living beings, sociocultural phenomena."[193]

Virilio argued that science and technology in the postmodern context are essentially the same. He stated that modern science has progressively become "technoscience," which is concerned with "immediate effectiveness" rather than truth.[194] Furthermore, he equated technoscience with technoculture and placed the responsibility for the acceleration of all reality squarely upon it.[195] In an assessment that brings to mind both Derrida's and Lyotard's characterizations of the postmodern ethos, he declared that technoscience is "concerned less with truth than the effect created by the announcement of a new discovery," regardless of whether it has any real value for the common good.[196]

Ethics, according to Virilio, have become irrelevant in postmodern technoscience, whose various disciplines are exiled "from all reason."[197] In his view, postmodern technoscience is "postscientific extremism," whose practitioners seek to achieve "limit performances" [i.e., go to the limit of the possible, regardless of the consequences] in fields like robotics and genetic engineering.[198] The "white-coated adventurer," he declared, "pushes himself to the ethical limits," experiencing "the elation of risking not only his own life, but that of the human race."[199]

The loss of normal points of reference with the new, digitally created reality. We have concentrated our attention in this work on transformations in the theory of technology and the dynamics over more than two millennia of philosophical ideas and moral principles that affected technological development and determined its importance in Western culture. In his analysis of postmodern technoculture, however, Virilio made the case that postmodern technoscience has undermined the ability of humans to experience the phenomenal world in realistic terms and to understand the consequences of their actions in it. Thanks to information technology and calculation, he declared, "[W]e are faced with the reconstruction of the phenomenology of perception according to the machine."[200] He lamented, "We are gradually deprived of our natural receptor organs," and are "obsessed . . . by a kind of cosmic lack of proper measure."[201] The human, he said, has ceased "to measure the world by his own scale" and is no longer "the measure of all things."[202]

Virilio claimed there is a growing erosion of natural human capabilities as technology itself progresses. "Each innovation," he argued, adds to a constellation of "detrital disorders," including those of a "visual, social, psychomotor, affective, intellectual, and sexual" character. They pile up as they are transmitted from generation to generation, not through the genes, but, according to Virilio, by means of "unutterable technical contamination" of the body and its perceptive and conceptual apparatus.[203]

He wrote of the cumulative effect of what Marshall McLuhan described over a generation earlier in *Understanding Media* as a process of "auto-amputation," by which individuals lose their natural capabilities as they come to rely on technologies to perform the same functions.[204] Thus, what for McLuhan was essentially a matter of habit was for Virilio a result of the pollution "of our sensory ecology once and for all."[205] Both suggested, in effect, that technological progress does not really "empower" individuals but in some important ways renders them weaker and more dependent on technology.

In Virilio's judgment, not only is there an accumulation of "specific injuries," but "all criticism of technology has just about disappeared." We are caught up, he lamented, in a "totalitarian techno-cult"—a web consisting of laws, social and moral strictures, and "what these centuries have made of us and our own bodies."[206] So, in addition to Ellul's theory that modern technology has become the object of quasi-religious worship, Marcuse's contention that a technology-based society is governed by "one-dimensional" positivist thought, and Postman's finding that a kind of cultural AIDS afflicts the information-glutted postmodern society, Virilio identified another reason—the impairment of our sensory and cognitive apparatus themselves by certain technologies—to explain the absence in technoculture of a viable critique or challenge to its dominance of life and society.

Our sense of time, a primary dimension of human experience, has been altered, according to Virilio, by global telecommunications and the blinding speed of transmission of digital information and images. Because of the phenomenon of "instantaneity," Virilio asserted, universal world time has replaced local time, the traditional basis of historical continuity. All of what we experience as reality is accelerated by the "light time" of virtual reality, he said, and everything is instantaneous and "now." The past, present, and future thus cede "to the immediacy of a tele-presence."[207] This phenomenon of speed creates a "feedback loop" and produces what Virilio called a "technological delirium," falsifying humans' understanding of the world around them and their relation to it.[208] Some even become addicted to the continual bombardment of images and symbols on the Web, according to Virilio, suffering what he termed "instant transmission sickness."[209]

At the same time that temporal intervals are shortened by the digital media, according to Virilio's analysis, space is dilated, creating a "planetary grand-scale optics." Visual continuity on a screen supplants the territorial contiguity of nations,[210] and a phenomenon of "delocalization" occurs. The "here" thus ceases to exist, Virilio maintained, and we are left only with the now.[211] In other words, postmodern technoculture is not only one in which there are no accepted metaphysical or moral points of reference and where "little narratives" are the essence of social and scientific discourse; it is also one where humans have lost their geographical footing in physical reality.

As spatial perception of the real world is transformed by digital technology, Virilio explained, humans are offered a substitute universe of experience—the virtual reality of cyberspace. The screen of a television set or computer is the new spatial frame of reference and, according to Virilio, replaces the "distant horizon line."[212] "A visual bubble of the collective imaginary," in which "the irrational . . . will flourish," seizes consciousness on a global basis.[213] This "progressive digitalization of audiovisual, tactile, and olfactory information" brings a "decline of immediate sensations," he said, and creates "the formidable threat of a . . . collective blindness on the part of humanity—the unprecedented possibility of the defeat of the facts."[214]

Suggesting that the creators of cyberspace wittingly encourage such a confusion of virtual images with reality, Virilio quoted the remarks of Microsoft's Bill Gates, who publicly declared he might "enjoy it" if in fact "the world existed only for him." For Virilio, these words reveal the "dimensional derangement" produced by digital technologies that scale the universe down to a "nursery . . . of the toys and games of an overgrown spoilt child."[215]

Virilio's total accident scenario, warning of a technological apocalypse. The progress of technoscience in the twentieth century, according to Virilio, creates a situation of extremes, where humanity is threatened by the danger of a "total," or global, accident, represented by the "three bombs"—information, nuclear, and genetic—"developing in parallel."[216] This danger exists, in his view, because "each time we invent a new technology, whether electronic or biogenic, we program a new catastrophe and an accident that we cannot imagine."[217] He referred to the nuclear, genetic, and cybernetic areas of technoscience as "bombs" not only because of their potential for unleashing catastrophic damage, but also to call attention to a "militarization of science," which he said has been going on for fifty years.[218]

"The three bombs," he added, ". . . reinforce one another. But in the middle . . . is the information bomb that is knowledge. It decides."[219] For Virilio, information technology is also a reason why the geographic scope of the accident could be "total." Since the speed of global information transmission is instantaneous, he warned, a worldwide catastrophe could be triggered by the dispatch of an improper command or erroneous information from a local point of origin.[220] Furthermore, he added, it could be "entirely impossible to distinguish a deliberate action from an involuntary reaction or an 'accident'; or to distinguish an attack from a mere technical breakdown."[221] He concluded, "To stem these tendencies . . . we can no longer fall back on a philosophical power or on a religious power in a broad sense of wisdom." He expressed the pessimistic view that humanity is now at a point where only critique remains.[222]

The Y2K scenario, according to which there would have been a global disruption in financial, military, air travel, medical, law enforcement, and water and power supply facilities due to a software error, was fortunately thwarted by billions of dollars of last-minute software modifications worldwide made at the end of the second Christian millennium. Virilio did not hesitate to cite it as an example, however, of the kind of global catastrophic phenomenon the information bomb could produce.[223] Virilio also identified as forerunners of "the Accident which will unfailingly bring down this house of cards" the 1987 stock market plunge in the United States, accelerated automatically downward by programmed trading, and the 1997 Asian financial crisis, spread by information technology from one nation to another and causing losses worth billions of dollars in a few hours.[224]

The explosion, fire, and partial meltdown of the Chernobyl nuclear reactor in the Ukraine in 1986 was an instructive example of a catastrophic accident concerning nuclear technology. The damage to people and property was calculated by the Soviets in the hundreds of billions of dollars, not to mention thousands of deaths from the fire and explosion, clean-up operations, and subsequent radiation illnesses. Although Virilio admitted that this accident was only "local," he pointed out that with cybernetics, such accidents could be "total."[225]

A likely scenario for something approaching a total accident with nuclear technology is an exchange of strategic nuclear missiles between nations resulting from erroneous information or a technological mishap. It must be remembered that in such circumstances, the computerization of command and control systems makes it possible to execute the launch order and complete the strike in just a few minutes. And as Virilio warned, the proliferation of nuclear weapons to more and more nations—India and Pakistan were the most recent cases he cited—makes such weapons "generally commonplace" and puts such an accident "on the agenda for the next millennium."[226]

The fact today that an autonomous terrorist group might secure and explode a nuclear device complicates the picture further by raising the possibility that a nation could misidentify the author of such an attack as a hostile nation and launch a strike against the nonaggressor in error. That nuclear warfare could bring "total" damage was made clear in 1983 at an international conference in Helsinki attended by five hundred scientists who warned that a limited exchange of nuclear missiles between the United States and the Soviet Union would have destroyed all or part of the ozone layer, exposing plants and animals to deadly ultraviolet radiation, and would have unleashed a nuclear-winter phenomenon, causing surface temperatures on the planet to drop to a level where most humans and animals would die.

An example of how the technology bombs "work in parallel," according to Virilio, is that the "information bomb is in the process of programming the genetic bomb [i.e., by sequencing the DNA of genes]."[227] Genetics, he said, "is about to replace atomic science and become the major science."[228] It is a good example, in his view, of the principle of the "double-bind," according to which a technology can be "the best of things" and the "worst of things" at the same time. Control over the book of nature promises the capacity to treat diseases, but, Virilio warned, genetic techniques will be "quickly used to treat people in perfect health."[229]

The implications of genetic technoscience are disturbing, he argued, because the "sciences threaten the species no longer (as in the past) by the radioactive destruction of the human environment, but by ... the control of the sources of life ..." [230] That is, "[T]he very principle of all individuated life" (i.e., respect for the genetic integrity of existing species) is now called into question.[231] Virilio thus sees genetic engineering as a manifestation of "that nihilism of an omnipotent progress which," he said, "runs through the twentieth century."[232] He added that the West's expansionist drives, having exhausted the geography of the earth, have been transferred "to the human body ... that last, still unexplored corner of the planet."[233] "The colonization of the space-time of living matter," Virilio declared, will be "the new frontier of the will to power of the technosciences."[234]

He referred to the human genome project as an inevitable basis for "cybernetic eugenicism," where "the single market demands the commercialization of the whole of living matter, the privatization of the genetic heritage of humanity."[235] He no doubt found confirmation of this view in the fact that some key supporters of a $3 billion bond measure in California to fund a stem cell research program in 2004 promoted it to the voters as a job-creating measure, good for the state's economy.[236]

In a query that brings to mind Professor Sloterdijk's appeal at the 1999 Bavaria Heidegger colloquium for a new anthropological technology to counter the "failure of humanism," Virilio pointedly asked, "Are we not right to suspect that experiments on the industrialization of living matter . . . will soon lead back to that old folly of the 'new man'?"[237] He warned of the "artificial selection of the human species," of a misguided quest for the "*transhuman*, built on the lines of transgenic crops, which are so much better [sic!] adapted to their environment."[238] This idea of using genetic technoscience to produce an "improved" version of humans, as well as the threat of "extinction" of existing species posed by biotechnological modification, led Virilio to evoke the memory of Auschwitz. There, Dr. Mengele performed his experiments on humans, and Jews were exterminated, in keeping with the Nazis' racist ideology and eugenics policy. "What is happening right now in genetics," Virilio declared, is a "continuation" of Auschwitz.[239]

He argued that the use of genetic technoscience for eugenic purposes would be a perversion of science; and, evoking his Christian roots, he suggested it would be a catastrophe in the moral, historical, and even theological sense.[240] "There is no genetic progress of being," he declared, ". . . no possibility of improving humanity by the techniques." Such efforts would only amount, in his view, to a catastrophe of science, to the death of science—what he termed the "Total Accident of science."[241] He added the qualification, however, that it would not mean the end of the world, but rather the end of *a* world, the world of humanity as we know it. "Humans are not central in the history of the universe," Virilio stated, citing "acquired knowledge in Judeo-Christianism and early philosophy" to make his point.[242]

*   *   *

It is our view that Virilio has correctly identified three forms of contemporary technoscience that represent a potential for catastrophe on a grand scale: nuclear science and technology, because of the threat of nuclear war and the danger of radioactive wastes; information technology, because of its ability to transmit globally and instantaneously erroneous information and commands; and genetic technoscience, because it can be used to tamper recklessly with the chemistry of "human nature," as well as the very basis of all organic life, without proper regard for broader moral, ecological, and biological implications.

One might question, however, whether Virilio has actually focused our attention on the manner in which the greatest harm from technology could occur. Yes, there have been technological accidents of a severe nature, and undoubtedly there will be many more. And yes, their incidence and magnitude will undoubtedly continue to escalate. On the other hand, is Virilio's concept of a total accident too apocalyptic, too fatalistic?

In the first place, an accident that is "total" would have to be final and global in its destructive effects. Outside the possibility of a nuclear winter phenomenon provoked by a nuclear war begun by accident, or perhaps a radical disruption of global precipitation and climate patterns caused by genetic pollution from ice-minus bacteria,[243] it is difficult to imagine an accident that would have a truly global effect. Furthermore, an accident is a single, one-time event, yet some of the most destructive contemporary technology-related phenomena, such as global pollution of the atmosphere, resource depletion, and species extinction, are the result of a gradual accumulation of deliberate actions by humans, who have ample reason to be aware of the consequences.

On a rhetorical level, Virilio's characterization of genetic, cybernetic, and nuclear technoscience as "bombs" no doubt correctly alludes to the important role of government-funded military research in their development. It also contributes to a contemporary brand of paranoia, however, by filling one's consciousness with more of the images of impending destruction that have become valuable political currency since the 9/11 terrorist attack. Do they feed an irresistible and masochistic temptation to indulge ourselves by accepting his apocalyptic brand of pessimism?

Virilio spoke of the prospect of a Total Accident in terms of inevitability. Indeed, the word "accident" implies a kind of powerlessness in the face of consequences which arrive despite the best of intentions. If these catastrophes are inevitable, and if we will not renounce the technologies that Virilio said will bring them, then perhaps there is nothing to be done. Perhaps fatalism—waiting for them to happen—is the only feasible alternative.

In adopting such a passive stance, however, we would be acquiescing in the quasi-theological determinism of Virilio, who stated that there is "no doubt" that history is following "the same course as the Apocalypse," except that "science plays a big part in it."[244] His words are an illustration, in any case, of how, in the short period of the last millennium, the theory of technology has gone from the idea of technological progress as recovery from the Fall to one of postmodern technoscience as bearer of the Apocalypse. Another way to think of the present situation, however, is in terms of the Apollonian goddess Nemesis, returning in the guise of global warming to punish humans for ignoring the principle of limits. Is it possible to restore them? We have to think that somehow humans will find the will and a way. This, in any case, is one of the great challenges of the twenty-first century.

Notes to Chapter 6

1. Jeremy Herf, *Reactionary Modernism: Technology, Politics, and Culture in Weimar and the Third Reich* (New York: Cambridge University Press, 1984), 19, 20.

2. Ibid., 39.

3. Ibid.

4. Ibid., x.

5. Ibid., 134.

6. Ibid., 130, 133, 143.

7. Ibid., 146.

8. Ibid., 38.

9. Ibid., 150.

10. Ibid., 60–61.

11. Ibid., 56–57, 59.

12. Ibid., 30.

13. Ibid., 66–67.

14. Ibid.. 67–68.

15. Ibid., 67.

16. Ibid., 58, 62–63.

17. Ibid., 52.

18. Ibid., 61–62.

19. Ortega y Gasset, "Thoughts on Technology," in *Philosophy and Technology*, ed. Carl Mitcham and Robert Mackey (New York: The Free Press, 1972), 300–1.

20. Herf, 29.

21. Ian Buruma, "The Anarch at Twilight," *New York Review of Books*, 24 June 1993, 27.

22. Herf, 86.

23. Ibid., 75.

24. Ibid., 102.

25. Ibid., 72.

26. Ibid., 77.

27. Arendt, *Totalitarianism*, 164.

28. Ibid., 96, 98, 102, 107, 116–17.

29. Ibid., 106.

30. Buruma, 27.

31. Herf, 94–5.

32. Ibid., 72.

33. Ibid., 72, 94.

34. Ibid., 77.

35. Ibid., 104.

36. Ibid.. 46.

37. Ibid., 47.

38. Albert Speer, *Inside the Third Reich* (New York: Macmillan Co., 1970), 523.

39. Herf, 201.

40. Ibid., 210.

41. Ibid., 204.

42. Ibid., 196.

43. Ibid., 194.

44. Ibid., 207.
45. Ibid., 206.
46. Speer, 524.
47. Ibid., 520.
48. Ibid., 521.
49. Herf, 112–13.
50. Ibid., 113–14.
51. James Miller, "Heidegger's Guilt," in *Salmagundi* (winter/spring 1996): 216.
52. Herf, 114.
53. Miller, 218–19, 226.
54. Herf, 115, 150.
55. Ibid., 113–4.
56. Miller, 229, 321.
57. See pages 271–2 of this work.
58. Heidegger, *The Question Concerning Technology*, 294, 298.
59. Ibid., 308–9.
60. Ibid., 296, 315–17.
61. See Friedrich Jünger, *The Failure of Technology* (Chicago: Henry Regnery Company, 1960).
62. See Horkheimer and Adorno, *The Dialectic of Enlightenment*. See also Horkheimer, *The Eclipse of Reason* (New York: Oxford University Press, 1947).
63. Neil Postman, *Technopoly* (New York: Vintage Books, 1993), 48.
64. See Paul Virilio, *The Information Bomb* (New York: Verso, 2000).
65. See Postman.
66. Ellul, *The Technological Society*, 78.
67. Jacques Ellul, "Technique and the Opening Chapters of Genesis," in *Theology and Technology*, 124–25, 133.
68. Ellul, *The Technological Society*, 428–29
69. Ellul, "The Relation between Man and Creation in the Bible," in *Theology and Technology*, 140, 147. "Technique and the Opening Chapters," in *Theology*, 134.
70. John Wilkinson, translator's introduction to *The Technological Society*, xv.
71. Ibid., 133–34.
72. Ibid., 21, 79.
73. Ibid., 79–80.
74. Ibid., 72–74.
75. Ibid., 103–4.
76. Ibid., 142, 144.
77. Ibid., 65, 77–78.
78. Ibid., 208–15.
79. Ibid., 211–12.
80. Ibid., 110–11.
81. Ibid.. 97, 99.
82. Ibid., 105–6.
83. Ibid., 99–100.
84. Ibid., 85–86.
85. Ibid., 135–40.
86. Herbert Marcuse. *One Dimensional Man* (Boston: Beacon Press, 1964), 154.
87. Ibid., 170–71.
88. Ibid., 180–81.

89. Ibid., 175.

90. Ibid., 172.

91. Ibid., 173, 184.

92. Ibid., 171, 179–80.

93. Ibid., 172.

94. Ibid., 17.

95. Ibid., 158.

96. Ibid., 56–57.

97. Ibid., 168–69.

98. Ibid., 151.

99. Ellul, *The Technological Society*, 145.

100. Ibid., 143.

101. See Noble, *The Religion of Technology*, Part 1.

102. Ellul, *The Technological Society*, 144–45.

103. Ibid., 145.

104. Ibid.

105. Ibid., 22.

106. Marcuse, 32.

107. Marcuse. *Eros and Civilization* (Boston: Beacon Press, 1966)l, 94.

108. Ellul, *The Technological Society, 115.*

109. Ibid., 377–80.

110. Ellul, *Propaganda* (New York: Alfred A. Knopf, 1965), 123–32.

111. Ibid., 25, 27.

112. Ibid., 254.

113. B. F. Skinner, *Beyond Freedom and Dignity* (New York: Bantam Books, 1971), 150

114. Ibid., 144–50.

115. Marcuse, *One Dimensional Man,* 19, 23, 48.

116. Marcuse, *Eros and Civilization*, 97–8.

117. Ibid., 99.

118. Ibid., 101.

119. Mark Lilla, "The Politics of Jacques Derrida," *New York Review of Books*, 25 June 1998, 37.

120. Ibid.

121. Ibid., 37.

122. Ibid.

123. Ibid.

124. Ibid.

125. Ibid., 38.

126. Ibid.

127. Aude Lancelin, "Cinq clés d'une pensée radicale," *Le Nouvel Observateur*, 14–20 Oct. 2004, 23.

128. Jean-François Lyotard, *The Postmodern Condition: A Report on Knowledge*, trans. Geoff Bennington and Brian Massumi (Minneapolis: University of Minnesota Press, 1991), 31–33, 35.

129. Ibid., 32–34.

130. Ibid., 46, 54.

131. Ibid, 44, 51.

132. Ibid., 3, 54.

133. Ibid., 41.

134. Samuel Enoch Stumpf, *Philosophy: History and Problems*, 3d ed. (New York: McGraw-Hill Book Company, 1983), 433.

135. Ibid., 432.

136. Lyotard, 40.

137. Ibid.

138. Ibid.

139. Ibid., 43.

140. Ibid., 9, 25, 65.

141. Ibid., 58–60.

142. Ibid., 60.

143. Fredric Jameson, forward to *The Postmodern Condition* (Minneapolis: Minnesota University Press,1984), xix.

144. Lilla, 38.

145. Lyotard, 60–61.

146. Ibid., 60.

147. Ibid., 66.

148. Ibid., 64–5.

149. Ibid., 65–6.

150. Jameson, xx.

151. Lyotard, 66–67.

152. Daniel Bell, *The Coming of Post-Industrial Society* (New York: Basic Books, 1973), 29–32.

153. Lyotard 51.

154. Ibid.

155. Andrew Feenberg, *Alternative Modernity* (Berkeley: University of California Press, 1995), 131.

156. Ibid., 128.

157. Ibid., 67.

158. Feenberg, 145–46.

159. Lilla, 36.

160. Postman, 71–72.

161. Ibid., 48.

162. Ibid., 111.

163. Ibid., 118.

164. Ibid., 112.

165. Ibid., 69.

166. Ibid., 70. 72.

167. Ibid., 63.

168. Ibid., 72–77.

169. Ibid., 70.

170. Ibid., 72.

171. Lyotard, 41.

172. Jeremy Rifkin, *The Biotech Century* (New York: Penguin Putnam Inc., 1999), 67.

173. Jeremy Rifkin, *Algeny* (New York: Penguin Books, 1983), 198.

174. In 2001, Celera Corp. estimated the human genome consisted of approximately 30,000 genes. *International Herald Tribune*, 14 Feb. 2001. In 2004, an international

consortium of researchers adjusted the figure to 20,000–25,000 genes. *International Herald Tribune*, 22 Oct. 2004.

175. Rifkin, *Algeny*, 190–96.

176. Rifkin, *Algeny*, 208–10, and *The Biotech Century*, 209.

177. Rifkin, *Algeny*, 228–29.

178. Ibid., 211–13.

179. Ibid., 238.

180. Ibid., 240.

181. Rifkin, *Algeny*, 207, 213, and *The Biotech Century,* 188–89, 215.

182. Rifkin, *The Biotech Century*, 247–50.

183. *The Amicus Journal* (spring 1993).

184. Rifkin, *The Biotech Century*, 74.

185. Ibid., 98.

186. *Le Monde*, 29 Sept. 1999.

187. Ibid.

188. Rifkin, *Algeny*, 232–3.

189. Stephen J. Gould; "Genetic Good News: Complexity and Accidents," in *International Herald Tribune*, 20 Feb. 2001.

190. *International Herald Tribune*, 14 Feb. 2001.

191. Paul Virilio and Sylvere Lotringer, *Crepuscular Dawn* (New York: Semiotext[e], 2002), 141.

192. *International Herald Tribune*, 6 April 2001.

193. Virilio, 115–17.

194. Ibid., 2,3.

195. Ibid., 3.

196. Ibid., 4.

197. Ibid., 2.

198. Ibid.

199. Ibid., 4.

200. Virilio and Lotringer, 141.

201. Virilio, 40.

202. Ibid., 104.

203. Ibid., 39.

204. Marshall McLuhan, *Understanding Media: The Extensions of Man* (New York: Signet Books, 1964), 52–53.

205. Virilio, 114.

206. Ibid., 39.

207. Ibid., 116–8.

208. Virilio and Lotringer, 150–1.

209. Virilio, 38.

210. Ibid., 12–3.

211. Ibid., 116, 120.

212. Ibid., 119.

213. Ibid., 112.

214. Ibid., 113–14.

215. Ibid., 93.

216. Ibid., 134–36. Virilio and Lotringer, 160–63.

217. Virilio and Lotringer, 146.

218. Ibid., 145.

219. Ibid., 136.
220. Ibid., 160. Virilio, 133–34.
221. Virilio, 141.
222. Virilio and Lotringer, 161.
223. Virilio, 133.
224. Ibid., 128–9.
225. Virilio and Lotringer, 160.
226. Virilio, 135.
227. Virilio and Lotringer, 136.
228. Ibid., 149.
229. Ibid., 156.
230. Virilio, 139–40.
231. Ibid., 140.
232. Ibid., 138.
233. Ibid., 55.
234. Ibid., 138.
235. Ibid., 132.
236. *International Herald Tribune*, 4 Nov. 4, 2004. See Also Connie Buck, Connie; "Hollywood Science," *The New Yorker*, 18 Oct. 2004.
237. Virilio, 136.
238. Ibid.
239. Virilio and Lotringer, 148.
240. Ibid.. 157–59.
241. Ibid., 159.
242. Ibid., 158–59.
243. Rifkin, *The Biotech Century*, 75–76.
244. Virilio and Lotringer, 154.

Bibliography

Abbott, Leonard D., ed. *Masterworks of Government*. Vol. 2. New York: McGraw-Hill Book Company, 1973.

Adorno, Theodor. *Aesthetic Theory*. Minneapolis: University of Minnesota Press, 1977.

Amicus Journal: Spring, 1993.

Arendt, Hannah. *The Life of the Mind: One Volume Edition*. New York: Harcourt Brace and Co., 1978.

————. *Totalitarianism*. New York: Harcourt, Brace, and World, Inc., 1968.

Aristotle. *The Pocket Aristotle*. Ed. Justin D. Kaplan. Trans. under the editorship of W. D. Ross. New York: Washington Square Press, 1961.

Augustine. *The City of God*. New York: Modern Library, 1950.

Bacon, Francis. *Essays, Advancement of Learning, New Atlantis, and Other Pieces*. Ed. Richard Foster Jones. New York: Odyssey Press, 1937.

Baudelaire, Charles. *Les Fleurs du Mal*. Paris: Garnier Frères, 1937.

Beatty, John. and Johnson, Oliver, eds. *Heritage of Western Civilization*. 6th ed. Vol. 2. Englewood Cliffs, NJ: Prentice Hall, 1987.

————. *Heritage of Western Civilization*. 7th ed. Vols. 1 & 2. Englewood Cliffs, NJ: Prentice Hall, 1991.

————. *Heritage of Western Civilization*. 8th ed. Vols. 1 & 2. Englewood Cliffs, NJ: Prentice Hall, 1995.

Beauvoir, Simone de. *The Ethics of Ambiguity*. New York: The Citadel Press, 1958.

————. *The Marquis de Sade: An Essay by Simone de Beauvoir, with Selections from His Writings*. New York: Grove Press, 1953.

Bell, Daniel. *The Coming of Post-Industrial Society*. New York: Basic Books, 1973.

Berkeley Daily Planet. 7 June 2000.

Bluhm, William T. *The Theories of the Political System*. Englewood Cliffs, NJ: Prentice Hall, Inc., 1965.

Brook, James. and Iain Boal, eds. *Resisting the Virtual Life: The Culture and Politics of Information*. San Francisco: City Lights, 1995.

Broome, J. H. *Pascal*. London: Edward Arnold, Ltd., 1965.

Buck, Connie. "Hollywood Science." *New Yorker*, 18 Oct. 2004.

Buruma, Ian. "The Anarch at Twilight." *The New York Review of Books*, 24 June 1993.

Burtt, E. A. *The Metaphysical Foundations of Modern Science*. Garden City, N.Y.: Doubleday Anchor Books, 1954.

Butterfield, Herbert. *The Origins of Modern Science*. Rev. ed. New York: The Free Press, 1957.

Camus, Albert. *L'Homme Revolté*. Paris: Gallimard. 1951.

———. *The Rebel*. New York: Vintage Books, 1963.

Cassirer, Ernst. *The Philosophy of the Enlightenment*. Boston: Beacon Press, 1955.

Chassang, A. and Charles Senninges, eds. *Recueil de textes littéraires français, XVII siècle.* Paris: Hachette, 1975.

Cicero. *De natura deorum: Academia*. Trans. H. Rockham. London: G. P. Putnam's Sons, 1933.

Cleve, Felix M. *The Giants of Pre-Sophistic Thought*. Vol. 2. The Hague: Martinus Nizov, 1969.

Commelin, P. *Mythologie Grecque Et Romaine*. Paris: Éditions Garnier Frères, 1960.

Commoner, Barry. *The Closing Circle*. New York: Bantam Books, 1972.

Condorcet, Nicolas de. *Esquisse d'un Tableau Historique des Progrès de l'Ésprit Humain,* with an introduction by Monique and Francis Hincker. Paris: Éditions Sociales, 1966.

Coppleston, Fredrick. *A History of Philosophy: Fichte to Nietzsche*. Vol. 7. Westminster, MD: The Newman Press, 1965.

Crepet, Jacques, Ed. *Les Plus Belles Pages de Charles Baudelaire*. Paris: Éditions Messen, 1950.

Darcos, Xavier, and Bernard Tartoyre, eds. *Le XVII siècle en lettres.* Paris: Hachette, 1987.

Descartes, René. *Discours de la Méthode*. Ed. M. Robert Derathé. Paris: Librarie Hachette, 1937.

Diderot, Denis. *Pensées philosophiques; Lettre sur les aveugles; Supplément au voyage de Bougainville*. Paris: Garnier-Flammarion, 1972.

Dostoevsky, Fedor. *The Short Novels of Dostoevsky*. Trans. Thomas Mann. New York: Dial Press, 1945.

Dyson, Freeman. "The Future Needs Us." *The New York Review of Books*, 3 Feb. 2003

Ebbenstein, William. *Great Political Thinkers*. 3[d] ed. New York: Holt, Rinehart and Winston. 1960.

Edman, Irwin, editor. *The Philosophy of Schopenhauer*. Introduction by editor. New York: The Modern Library, 1928.

Edwards, Paul, ed. *The Encyclopedia of Philosophy*. Vols. 2, 3, 4, & 5. New York: Macmillan Company and The Free Press, 1967, 1969.

Ellul, Jacques. *The Political Illusion*. New York: Alfred A. Knopf, 1967.

————. *Propaganda*. New York: Alfred A. Knopf, 1965.

————. "The Relation between Man and Creation in the Bible." In *Theology and Technology: Essays in Christian Analysis and Exegesis*. Ed. Carl Mitcham and Jim Grote. Lanham, MD: University Press of America, 1984.

————. "Technique and the Opening Chapters of Genesis." In *Theology and Technology: Essays in Christian Analysis and Exegesis*. Ed. Carl Mitcham and Jim Grote. Lanam, MD: University Press of America, 1984.

————. *The Technological Society*, trans. New York: Vintage Books, 1964.

Feenberg, Andrew. *Alternative Modernity*. Berkeley: University of California Press, 1995.

Feuer, Lewis. "The Scientific Intellectual." In *The Psychological and Sociological Origins of Modern Science*. New York: Basic Books, 1963.

————. *Marx and the Intellectuals*. Garden City, NY: Anchor Books, 1969.

Fichte, Johann G. *Johann Gottlieb Fichte's Popular Works*, with a Memoir by William Smith, L.L.S. London: Trübner and Co., 1873.

————. *The Vocation of Man*. Chicago: Open Court Publishing Co., 1910.

Florman, Samuel. "In Praise of Technology." *Harpers*, Nov. 1975.

Fourier, Charles. *The Utopian Vision of Charles Fourier*. Ed. Jonathan Beecher and Richard Bienvenue. Boston: Beacon Press, 1971.

Freud, Sigmund. *Civilization and Its Discontents*. Trans. James Strachey. New York: W. W. Norton, 1961.

Fromm, Erich. *Marx's Concept of Man*. New York: Fredrick Ungar Publishing Co., 1969.

Fuentes, Carlos. "Writing in Time." *Explorations*, Jan. 1982.

Galgan, Gerald J. *The Logic of Modernity*. New York: New York University Press, 1982.

Gay, Peter. *The Enlightenment: An Interpretation*. Vols. 1 & 2. New York: Alfred A. Knopf; 1967, 1969.

Glacken, Clarence J. *Traces on the Rhodian Shore*. Berkeley: University of California Press, 1967.

Goethe, Johann Wolfgang von. *Faust: A Tragedy*. New York: Modern Library, 1967.

Goldstein, Thomas. *Dawn of Modern Science*. Boston: Houghton Mifflin, 1980.

Grant, Michael. *From Alexander to Cleopatra: The Hellenistic World*. New York: The Macmillan Publishing Company, 1990.

Grene, David, and Richard Lattimore, eds. *Greek Tragedies*. Vol. 1. Chicago: University of Chicago Press, 1960.

Gould, Stephen J. "Genetic Good News: Complexity and Accidents." *International Herald Tribune*, 20 Feb. 2001.

Habermas, Jürgen. *The Philosophical Discourse of Modernity*. Cambridge: MIT Press, 1987.

Hahn, Roger. "Laplace and the Mechanical Universe." In *God and Nature: Historical Essays on the Encounter between Christianity and Science.* Berkeley: University of California Press, 1986.

————. "Laplace and the Vanishing Role of God." In *The Analytical Spirit: Essays in the History of Science in Honor of Henry Guerlac.* Ed. Harry Woolf. Ithaca: Cornell University Press, 1981.

Hayden, Hiram. *The Counter-Renaissance.* New York: Charles Scribner's Sons, 1950.

Hegel, G. W. F. *Phenomenology of Spirit.* Trans. A. V. Miller. New York: Oxford University Press, 1977.

Heidegger, Martin. *The Question Concerning Technology and Other Essays.* New York: Harper Torchbacks, 1977.

Heisenberg, Werner. *Physics and Philosophy: The Revolution in Modern Science.* New York: Harper Torchbacks, 1962.

Heller, Eric. *The Disinherited Mind.* New York: Farrar, Straus, Giroux, 1957.

Herf, Jeremy. *Reactionary Modernism: Technology, Culture, and Politics in Weimar and the Third Reich.* New York: Cambridge University Press, 1984.

Hollingsdale, R. J. *Friedrich Nietzsche: The Man and His Philosophy.* Baton Rouge: LSU Press, 1965.

Holy Bible. The New King James Version. New York: Thomas Nelson Publishers, 1982.

Horkheimer, Max. *The Eclipse of Reason.* New York: Oxford University Press, 1947.

Horkheimer, Max, and Theodor Adorno. *The Dialectic of Enlightenment.* New York: Continuum Publishing Company, 1972.

Hume, David. *Dialogues Concerning Natural Religion.* New York: Hafner Press, 1948.

Hutin, Serge. *L'Alchimie.* Paris: Presses Universitaires de France, 1971.

International Herald Tribune, 30 Jan. 2001, 14 Feb. 2001, 6 April 2001, 7 May 2001, 10 June 2001, 31 July 2001, 4 Nov. 2004.

Jaeger, Werner. *Paideia: The Ideals of Greek Culture.* Vol. 1. New York: Oxford University Press, 1967.

Joy, Bill. "Why the Future Doesn't Need Us." *Wired* 8.04 (April 2000).

Jünger, Friedrich. *The Failure of Technology.* Chicago: Henry Regnery Company, 1960.

Kant, Immanuel. *The Philosophy of Kant: Immanuel Kant's Moral and Political Writings.* New York: The Modern Library, 1949.

Kaufman, Walter A. *Nietzsche: Philosopher, Psychologist, Antichrist.* Princeton: Princeton University Press, 1950.

Kolakowski, Leszek. *The Alienation of Reason.* Garden City, NY: Anchor Books, 1969.

Koyré, Alexandre. *From the Classical World to the Infinite Universe.* Baltimore: Johns Hopkins Press, 1957.

Lagarde, André, and Laurent Michard, eds. *XVIII Siècle: Les Grands Auteurs de Programme.* Paris: Éditions Bordas, 1970.

————. *XVII Siècle: Les Grands Auteurs de Programme.* Paris: Éditions Bordas, 1953.

Lame Deer, John Free, and Richard Erodes. *Lame Deer, Seeker of Visions*. New York: Simon and Schuster, 1972.

Lancelin, Aude. "Cinq clés d'une pensée radicale." *Le Nouvel Observateur*, 14–20 Oct. 2004.

Lang, Helen S. *Aristotle's* Physics *and Its Medieval Varieties*. Albany, NY: State University of New York Press, 1992.

Leiss, William. *The Domination of Nature*. New York: George Braziller, 1972.

Le Monde, 29 Sept. 1999.

Lévi-Strauss, Claude. *La Pensée Sauvage*. Paris: Librarie Plon, 1962.

Lilla, Mark. "The Politics of Jacques Derrida." *The New York Review of Books*, 25 June 1998.

Lindberg, C., and Ronald C. Numbers, Ed. *God and Nature: Historical Essays on the Encounter Between Christianity and Science*. Berkeley: University of California Press, 1986.

Lovejoy, Arthur O. *The Great Chain of Being*. Cambridge, MA: Harvard University Press, 1978.

Lyotard, Jean-François. *The Postmodern Condition: A Report on Knowledge*. Trans. Geoff Bennington and Brian Massumi, with a forward by Fredric Jameson. Minneapolis: University of Minnesota Press, 1984.

Mailer, Norman. *Of a Fire on the Moon*. New York: The New American Library, 1971.

Mandelbaum, Maurice. *History, Man, and Reason: A Study in Nineteenth Century Thought*. Baltimore: Johns Hopkins Press, 1975.

Manuel, Frank. *The Prophets of Paris*. New York: Harper Torchbooks, 1962.

———. *The Religion of Isaac Newton*. Oxford at the Clarendon Press, 1974.

Marcuse, Herbert. *Eros and Civilization*. Boston: Beacon Press, 1966.

———. *One Dimensional-Man*. Boston: Beacon Press, 1964.

Marx, Karl. *Karl Marx, Selected Writings*. Ed. David McClellan. London: Oxford University Press, 1977.

Marx, Karl, and Friedrich Engels. *Karl Marx and Friedrich Engels: Selected Works in One Volume*. Ed Progress Publishers, Moscow. London: Lawrence and Wishart, 1968.

———. *The Marx-Engels Reader*. Ed. Robert Tucker. New York: W. W. Norton and Co., Inc., 1972, 1978.

Means, Russell. "Fighting Words on the Future of the Earth." In *Questioning Technology*. Ed. John Zerzan and Alice Carnes. Philadelphia: New Society Publishers, 1991.

Mendel, Arthur P., ed. *The Essential Works of Marxism*. New York: Bantam Books, 1963.

McKibben, Bill. "Bush in the Greenhouse." *The New York Review of Books*, 5 July 2001.

McLuhan, Marshall. *Understanding Media: The Extensions of Man*. New York: Signet Books, 1964.

Merchant, Carolyn. *The Death of Nature*. San Francisco: HarperSanFrancisco, 1989.

Merton, Robert K. *Science, Technology, and Society in Seventeenth-Century England.* New York: Harper Torchbacks, 1970.

Miller, James. "Heidegger's Guilt." *Salmagundi* (winter/spring 1996).

Moss, Norman. *Men Who Play God.* Baltimore, MD: Penguin Books, Inc., 1972.

Mumford, Lewis. *The Myth of the Machine: Technics and Human Development.* Vol. 1. New York: Harcourt Brace Jovanovich, Inc., 1966.

———. *Technics and Civilization.* New York: Harcourt, Brace and World, Inc., 1963.

New York Times, 5 Jan. 2001.

Newton, Isaac. *Newton's Mathematical Principles.* Berkeley: University of California Press, 1946.

———. *Opticks.* New York: Dover Publications, 1952.

Nietzsche, Friedrich. *The Birth of Tragedy and The Case of Wagner.* New York: Vintage Books, 1967.

———. *The Joyful Wisdom.* New York: Fredrick Ungar Publishing Co., 1960.

———. *The Portable Nietzsche.* Ed. Walter A. Kaufman. New York: The Viking Press, 1954.

———. *The Will to Power.* Ed. Walter Kaufman. New York: Vintage Books, 1968.

Nisbet, Robert. *The History of the Idea of Progress.* New York: Basic Books, 1980.

Noble, David. *The Religion of Technology.* New York: Alfred A. Knopf, 1998.

Ortega y Gasset, José. *Meditación de la técnica y otros ensayos sobre ciencia y filosofía.* Madrid: Revista de Occidente en Alianza Editorial, 1982.

———. "Thoughts on Technology." In *Philosophy and Technology.* Ed. Carl Mitcham and Robert Mackey. New York: The Free Press, 1972.

Paglia, Camille. *Sexual Personae.* New York: Vintage Books, 1991.

Pascal, Blaise. *Pensées.* Paris: Hachette, 1950.

Plato. *The Collected Dialogues of Plato.* Ed. Edith Hamilton and Huntington Cairns. Princeton: Princeton University Press, 1953.

———. *The Dialogues of Plato.* Vol. 2. Trans. B. Jowett. New York: Random House, 1892.

———. *The Republic of Plato.* Trans. Francis Cornford. London: Oxford University Press, 1941.

Platon. *Oeuvres complètes.* Vol 1. Trans. and notes, Léon Robin. Paris: Éditions Gallimard, 1950.

Postman, Neil. *Technopoly.* New York: Vintage Books, 1993.

Prosch, Harry. *The Genesis of Twentieth-Century Philosophy.* Garden City, N.Y.: Anchor Books, 1966.

Rather, L. J. *The Dream of Self-Destruction.* Baton Rouge: LSU Press, 1979.

Reik, Miriam M. *The Golden Lands of Thomas Hobbes.* Detroit: Wayne University Press, 1977.

Rifkin, Jeremy. *Algeny.* New York: Penguin Books, 1983.

———. *The Biotech Century.* New York: Penguin Putnam Inc., 1999.

Rosen, Charles, and Henri Zerner. "What Is, and Is Not, Realism." *The New York Review of Books*, 18 Feb. 1982.

————. "Enemies of Realism." *The New York Review of Books*, 4 Mar. 1982.

Rosen, Stanley. *Nihilism: A Philosophical Essay.* New Haven and London: Yale University Press, 1969.

Roszak, Theodore. *Where the Wasteland Ends.* New York: Anchor Books, 1973.

Rousseau, Jean-Jacques. *Dialogues; Rêveries d'un Promeneur Solitaire (Extraits).* Paris: Librarie Larousse, 1941.

————. *Discours sur les Sciences et Les Arts; Discours sur l'Origine et les Fondements de l'Inégalité parmi les Hommes.* Paris: Garnier-Flammarion, 1971.

Russell, Bertrand. *A History of Western Philosophy.* New York: Simon and Schuster, 1972.

Sabine, George H. *A History of Political Theory.* New York: Holt, Rinehardt, and Winston; 1961.

Santayana, George. *The German Mind.* New York: Thomas Crowell, 1968.

Schiller, Friedrich. *On The Aesthetic Education of Man.* New Haven: Yale University Press, 1954.

Schwartz, Eugene. *Overskill.* New York: Ballantine Books, 1971.

Sebestik, Jan. "The Beginning of Technological Thinking in the Late Eighteenth and Early Nineteenth Centuries." Text of a lecture given on a visit from L'Institut d'Histoire des Sciences, Paris, at the University of California, Berkeley, 1 Dec. 1981.

Sills, David L., Ed. *International Encyclopedia of Social Science.* Vol 9. New York: Macmillan Co. and The Free Press, 1968.

Skinner, B. F. *Beyond Freedom and Dignity.* New York: Bantam Books, 1971.

Speer, Albert. *Inside the Third Reich.* New York: Macmillan Co., 1970.

Spink, J. S. *French Free Thought, from Gassendi to Voltaire.* University of London: The Athlone Press, 1960.

Starkie, Enid. *Baudelaire.* London: Victor Gallancz Ltd., 1957.

————. *Flaubert: The Making of a Master.* New York: Atheneum, 1967.

Strong, Tracy B. *Friedrich Nietzsche and the Politics of Transfiguration.* Berkeley: University of California Press, 1975.

Stumpf, Samuel Enoch. *Philosophy: History and Problems.* 3^d ed. New York: McGraw-Hill Book Company, 1983.

Swift, Jonathan. *Gulliver's Travels.* New York: Rand McNally and Company, 1941.

Terdiman, Richard (1976). *The Dialectic of Isolation.* New Haven: Yale University Press, 1967.

Tuveson, Ernest L. *Millenium and Utopia.* Berkeley: University of California Press, 1949.

Van Deusen, Neil. *Telesio: The First of the Moderns.* Ph.D. diss., Columbia University, 1932.

Virilio, Paul. *The Information Bomb*. Trans. Chris Turner. New York: Verso, 2000.

Virilio, Paul, and Sylvère Lotringer. *Crepuscular Dawn*. Trans. Mike Taormina. New York: Semiotext(e), 2002.

Wall Street Journal, 17 Sept. 1986, 1 June 1992, 29 Sept.1999.

White, Lynn Jr. *Medieval Technology and Social Change*. New York: Oxford University Press, 1962.

———. "The Historical Roots of Our Ecological Crisis." *Science*, 10 Mar. 1967.

Whitney, Elspeth. *The Mechanical Arts in the Context of Twelfth- and Thirteenth-Century Thought*. Ann Arbor: University Microfilms International, 1985.

Winner, Langdon. *The Whale and the Reactor: A Search for Limits in the Age of High Technology*. Chicago: University of Chicago Press, 1986.

Yates, Francis. "Science, Salvation, and the Cabala." *The New York Review of Books*, 27 May 1976.

Zerzan, John. Article in *Fifth Estate* 20, no. 2 (1985).

Zimmerman, Michael. "Current Debate: Nature and Domination." *Tikkun* 4, no. 2 (1989).

Index

About the Author

Gregory Davis was born in San Francisco in 1934, and educated at Stanford University and the University of Aix-Marseille, France. He currently lives in Berkeley, California. During more than thirty years' residence in Silicon Valley, he was constantly exposed to a prevailing belief that technology and technological modes of thinking represented the best prospect for humanity. His own view on technology was colored, however, by exposure to military technology while serving in the U.S. Navy in the Far East in the late 1950s, employment from 1962–64 as a Political Scientist at Stanford Research Institute investigating strategic and political questions concerning nuclear weapons, a growing perception of dehumanization and environmental degradation caused by technology, and an encounter with Jacques Ellul's powerful analysis of modern technology, *The Technological Society*. Davis, in effect, became convinced that technology was not only a determining factor in contemporary life but also an increasingly problematical one. In 1973, he inaugurated the course Technology, Contemporary Society, and Human Values, which he taught for the next twenty-seven years at the College of San Mateo. He also taught for many years at the college a course on political theory and a colloquium on the history of Western ideas. In 1981, University Press of America published his book, *Technology—Humanism or Nihilism*, which was translated into Spanish and published in 1984 by Edamex in Mexico as *¿Tecnología: Esclavitud o liberación?* The present work, *Means Without End*, is the result of a project conceived when Davis was a research fellow at the Office of the History of Science and Technology at the University of California, Berkeley.